NIMS Monographs

The NIMS Monographs are published by the National Institute for Materials Science (NIMS), a leading public research institute in materials science in Japan, in collaboration with Springer. The series present research results achieved by NIMS researchers through their studies on materials science as well as current scientific and technological trends in those research fields.

These monographs provide readers up-to-date and comprehensive knowledge about fundamental theories and principles of materials science as well as practical technological knowledge about materials synthesis and applications.

With their practical case studies the monographs in this series will be particularly useful to newcomers to the field of materials science and to scientists and engineers working in universities, industrial research laboratories, and public research institutes. These monographs will be also available for textbooks for graduate students.

National Institute for Materials Science
http://www.nims.go.jp/

More information about this series at http://www.springer.com/series/11599

Toshihiro Hanamura · Hai Qiu

Analysis of Fracture Toughness Mechanism in Ultra-fine-grained Steels

The Effect of the Treatment Developed in NIMS

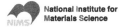

National Institute for
Materials Science

Springer

Toshihiro Hanamura
Hai Qiu
National Institute for Materials Science
Tsukuba
Japan

ISSN 2197-8891 ISSN 2197-9502 (electronic)
ISBN 978-4-431-54498-2 ISBN 978-4-431-54499-9 (eBook)
DOI 10.1007/978-4-431-54499-9

Library of Congress Control Number: 2014946962

Springer Tokyo Heidelberg New York Dordrecht London

Printed on acid-free paper

Springer is part of Springer Science+Business Media (www.springer.com)

Preface

The National Institute for Materials Science (NIMS) of Japan has long been working on research and development projects related to advanced steels, and now the development of environmentally friendly steel products has become an urgent necessity. The main idea in the production-process project is to utilize technologies for refined and ultra-refined steels. Mechanical properties of ultra-grain refined steels are almost equivalent to those of conventional high-strength and high-toughness steels produced with rare and expensive alloying elements. In this book, advanced steels developed at NIMS, based primarily on unique structural controlling process techniques, their mechanical properties, and their property improvement mechanisms are presented and discussed. The following is an overview of each chapter.

Chapter 1 Introduction: Environmental Problems and Related Advanced Steel Techniques (T. Hanamura)

Tensile strength and toughness of advanced steels are discussed from both the fundamental and engineering points of view.

Chapter 2 Ultra-Fine-Grained Steel: Relationship Between Grain Size and Tensile Properties (T. Hanamura)

Some phenomena in tensile properties unique to ultra-fine grained steels are presented and discussed. These phenomena include an increase in the strain-hardening rate with increasing carbon content up to 0.3 wt. % C and a strain-hardening rate almost constant even with a further increase in carbon content.

Chapter 3 Ultra-Fine-Grained Steel: Relationship Between Grain Size and Impact Properties (T. Hanamura)

Impact properties unique to ultra-fine grained steels are discussed. These properties include excellent fracture toughness owing to the steel's characteristic small effective grain size and high surface energy of fracture in comparison with those of other steel structures.

Chapter 4 Ultra-Fine-Grained Steel: Fracture Toughness (Crack-Tip-Opening Displacement) (H. Qiu)

The variation tendency of fracture toughness with ferrite grain size is discussed in detail, and the fracture toughness of ultra-fine grained steel is evaluated.

Chapter 5 Summary (T. Hanamura)

The main focus of the book is summarized, namely the tensile strength and toughness of advanced steels from both the fundamental and engineering points of view.

Acknowledgments

The authors acknowledge experimental support extended by Satoshi Iwasaki, Koji Nakazato, Takaaki Hibaru, Syuji Kuroda, Yasushi Taniuchi, Sadao Hiraide, Masayuki Komatsu, Goro Arakane, Masahiko Kawasaki, Noboru Sakurai, Elena Bulgarevich, Aiko Takanabe, and Mai Tadono who have immensely helped in the construction of this text book.

Contents

Chapter 1
Introduction: Issues Concerning Environmental Problems and Related Advanced Steel Techniques

Environmental problems have gained focus globally; therefore, it has become increasingly necessary to develop environment-friendly steel products. The amount of steel scraps produced each year in Japan has been predicted to increase constantly and to eventually surpass the total amount of steels annually produced in Japan. From an environmental point of view, therefore, it has become important to use increasing amounts of domestic steel scraps for steel production for the sustainable growth of Japan as well as the world.

With these environmental concerns in mind, the National Institute for Materials Science, (NIMS) has long been working on research and development projects concerning advanced steels and their production techniques. One of the aims of these projects is to produce steels with simple alloying elements such as C, Mn, and Si. This is aimed at systematically making use of steel scraps in recycling processes, whose amount is expected to drastically increase in the near future. This is hopefully expected to lead to steel-to-steel recycling process technologies, which might make it possible to achieve advanced recycling processes such as car-to-car recycling and help replace the current conventional so-called cascade-type recycling (which degrades product quality with an increase in the number of cycles). The main idea in developing steel-production processes by using steel scraps is to produce refined/ultra-refined grain-structural steels with mechanical properties almost equivalent to those of conventional high-strength and high-toughness steels produced by incorporating of some rare alloying elements such as Ni, Cr, V, Mb and Mo.

This book presents and discusses the advanced steels developed at NIMS [1–7] by primarily using unique structure-controlling techniques, their mechanical properties and their property improvement mechanisms are to be presented and discussed. The main focus is, in particular, on total balance between the tensile strength properties and fracture toughness properties in steels, from fundamental as well as engineering points of view. For fundamental studies, a unique analysis of the fracture surface energy in relation to the effective grain size (i.e., the unit size of a brittle fracture surface, which in most cases corresponds to the material's grain

© National Institute for Materials Science, Japan. Published by Springer Japan 2014
T. Hanamura and H. Qiu, *Analysis of Fracture Toughness Mechanism in Ultra-fine-grained Steels*, NIMS Monographs, DOI 10.1007/978-4-431-54499-9_1

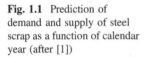

Fig. 1.1 Prediction of demand and supply of steel scrap as a function of calendar year (after [1])

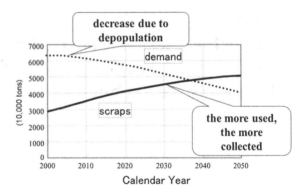

size) is employed to better understand the property improvement mechanisms and the characteristics of ultra-fine-grained steels. However, in engineering studies, fracture toughness parameters such as values of crack-tip-opening displacement (CTOD) in advanced steels needs to be evaluated and compared with those of conventional commercial steels. Therefore, this unique analysis would prove to be very helpful for industry-based researchers and engineers, as well as for university graduate students to get acquainted with some of the current advances in steel technologies, specifically with regard to the optimization of the total balance of mechanical properties in recently developed advanced steels for environment-friendly applications.

A prediction of the change in the demand and supply of steel scraps with time is presented in Fig. 1.1. It is interesting to note that in Japan, the demand for steel is predicted to gradually decrease, while the total amount of steel scraps is predicted to drastically increase. This prediction is mainly attributed to the fact that the domestic infrastructure of Japan developed since the end of the World War II has been constantly improving with time, while the population of the country has been decreasing because of the continued decrease in the number of newly born babies and increase in the average longevity. This indicates the possibility that Japan will eventually become self-sufficient in steels and will no longer require any further import of iron ores from countries such as Australia and Brazil for the domestic steel production. From Fig. 1.1 it is also possible to predict that around 2,037, the total amount of annual steel demand and annually produced steel scraps would eventually become equal to each other. This is expected to be the time when Japan would become a country that does not require iron import. Hence, Japan can eventually gain the ability to supply steel resources to herself without depending on outside resources.

However, the change in scrap steel prices is quite unpredictable, as can be seen from Fig. 1.2. The price of scrap steel often changes rapidly, and this tendency coincides somehow roughly with the change in the price of small steel rods. This drastic change in the prices of scraps and steels has a tremendous influence on the

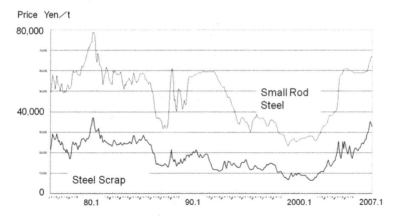

Fig. 1.2 Change in scrap steel price (after [1])

managements of scrap industries; therefore, it may be difficult to maintain the supply of iron resources stably. However, if Japan can become free from importing iron resources and manage steel production on its own, it is much more flexible and more stable to maintain a constant supply of steels without being strongly affected by unpredictable changes in the global economy.

However, if scrap steels are used as iron resources, there will be other difficulties that Japan will be forced to deal with. One is the contamination of steel products by repeated recycling. Scrap steels generally contain impurities originating from sources such as copper wires in automobiles and tin or zinc coatings in canned products. Another source of contamination is the presence of different types of alloying elements. Certain alloying elements are added to individual steels to impart special properties to steel products, as shown in Table 1.1.

In Table 1.1, the gray region indicates amounts of some alloying elements included in each of the commercial steels mentioned in the table. In many cases, these alloying elements are necessary for improving the mechanical properties of steel products, such as strength, ductility, toughness, and corrosion resistance. There are several methods for increasing the strength of steel; among them, the addition of alloying elements is rather a convenient method that can easily achieve strengthening through either solution hardening or precipitation hardening as the strengthening mechanism. However, the addition of alloying elements has its disadvantage; in most cases, the increase in the strength often decreases the ductility and toughness of steel products. At the same time, it is also possible to increase the strength by grain refinement. In this case, addition of extra alloying elements to the steel is not necessary, whereas only thermo-mechanical treatment is necessary to achieve strengthening equivalent to that achieved through addition of alloying elements. Furthermore, it is also possible to increase the fracture toughness by grain refinement. This technique of simultaneously improving strength and toughness is, therefore, environment-friendly as well as good for recycling of steels, because of the absence of additional contamination due to the inclusion of extra alloying

Table 1.1 Chemical composition of commercial steels

No.	Steel	C	Si	Mn	P	Si	Cu
800–1,000 MPa steel							
1	WEL–TEN80	≤0.18	0.15–0.35	0.60–1.20	≤0.035	≤0.040	0.15–0.50
2	HE	≤0.18	0.15–0.35	0.60–1.20	≤0.030	≤0.030	0.15–0.50
3	River Ace K–O	≤0.18	≤0.35	≤1.00	≤0.030	≤0.030	≤0.50
4	NK-HiTEN80	≤0.18	0.15–0.35	≤1.00	≤0.035	≤0.040	0.15–0.50
5	WEL-CON2HUltra	0.08–0.16	≤0.55	0.60–1.20	≤0.035	≤0.040	0.15–0.50
6	WEL-TEN100	≤0.18	0.15–0.35	0.60–1.20	≤0.035	≤0.040	0.15–0.50

No	Ni	Cr	Mo	V	B	YS (MPa)	TS (MPa)	TE (%)
1	≤1 50	0.40–0.80	≤0.60	≤0.10	≤0.006	≤700	800–950	≥18
2	0.70–1.00	0.40–0.80	0.40–0.60	0 03–0 10	0.002–0.006	≤700	800–950	≥18
3	≤1.50	≤0.80	≤0.60	<0.08	≤0.006	≤700	800–950	≥18
4	≤1.00	≤0.80	≤0.60	≤0.10	≤0.006	≤700	≥800	≥18
5	≤1.50	≤0.80	≤0.70	≤0.10	≤0.006	≤700	800–950	≥18
6	≤1.50	0.40–0.80	≤0.60	≤0.10	–	≥900	970–115	≥15

Fig. 1.3 Estimation of yield strength and DBTT (ductile-brittle transition temperature based on Pickering's equations (after [1])

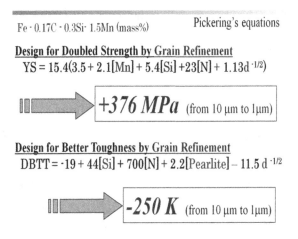

Fe · 0.17C · 0.3Si· 1.5Mn (mass%)   Pickering's equations

Design for Doubled Strength by Grain Refinement
$$YS = 15.4(3.5 + 2.1[Mn] + 5.4[Si] + 23[N] + 1.13d^{-1/2})$$

$+376\ MPa$ (from 10 μm to 1μm)

Design for Better Toughness by Grain Refinement
$$DBTT = -19 + 44[Si] + 700[N] + 2.2[Pearlite] - 11.5\ d^{-1/2}$$

$-250\ K$ (from 10 μm to 1μm)

elements, although recycling with the same steel composition is also possible. This can eventually achieve a sustainable environment-friendly society.

In predicting the total balance of steel properties, some empirical equations for predicting yield strength (YS) and ductile-to-brittle transition temperature (DBTT) of steel have been presented by Pickering [8], and through these equations it is possible to estimate the strength of steels. Some of the Pickering's equations are presented in Fig. 1.3 which also lists the factors determining the strength and DBTT, including the contents of Mn, Si, N, pearlite volume fraction, and grain size. It is also quite interesting to note that grain size significantly affects the strength as a function of $d^{-1/2}$, where d is the average grain size of steel. Hence, when the grain size is decreased from 10 to 1 μm, the strength is predicted to increase by 375 MPa and the DBTT is predicted to decrease by 250 K. Images illustrating ultra-fine-grained steel production processes are shown in Fig. 1.4, and an outline of processes involved in ultra-fine-grained steel production is given in Fig. 1.5. These figures show that it is quite effective to provide extensive strain from different directions to steel in order to produce an ultra-fine-grained steel. The remarkable performance of the ultra-fine-grained steel is shown in Figs. 1.6 and 1.7. A comparison of the stress-strain curves of the original 400 MPa steel with the conventional 800 MPa steel and ultra-fine-grained steel is shown in Fig. 1.6. A comparison of the brittle surface fraction in the Charpy impact test between the conventional 800 MPa steel and ultra-fine-grained steel is shown in Fig. 1.7. As can

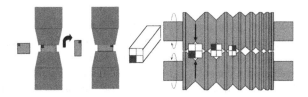

Fig. 1.4 Schematic figures of ultra-fine-grained steel production process (after [1])

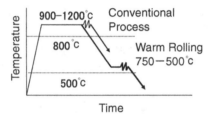

Fig. 1.5 Outline of the ultra-fine-grained steel production process (after [1])

Fig. 1.6 Comparison of stress-strain curves between original steel, conventional 800 MPa steel, and ultra-fine-grained steel (after [1])

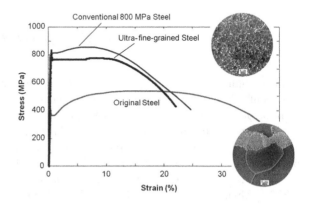

Fig. 1.7 Comparison of brittle surface fraction in Charpy impact test between conventional 800 MPa steel and ultra-fine-grained steel (after [1])

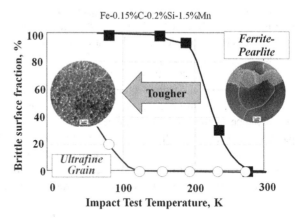

be seen in these figures, ultra-grain refinement is more advantageous for simultaneously improving strength and DBTT, compared to conventional steels.

On the other hand, in comparison to the grain refinement effect, other strengthening mechanisms such as solid solution hardening, precipitation hardening and work hardening are disadvantageous because although these mechanisms increase the tensile strength (TS), the DBTT also increases simultaneously.

Therefore, among all strengthening mechanisms, grain refining is the only method to simultaneously improve TS and DBTT.

The mechanism that is involved in the improvement of mechanical properties through grain refinement is one of the main topics of this book.

References

1. K. Nagai, Ultrafine-grained steels: basic researches and attempts for application. Can. Metall. Q. **44**(2), 187–194 (2005)
2. S. Torizuka, A. Ohmori, S.V.S.N. Murty, K. Nagai, Effect of strain on the microstructure and mechanical properties of multi-pass warm caliber rolled low carbon steel. Scripta Mater. **54**, 563–568 (2006)
3. A. Ohmori, S. Torizuka, K. Nagai, N. Koseki, Y. Kogo, Effect of deformation temperature and strain rate on evolution of ultrafine-grained structure through single-pass large-strain warm deformation in a low carbon steel. Mater. Trans. **45**(7), 2224–2231 (2004)
4. M. Zhao, T. Hanamura, H. Qui, K. Nagai, K. Yang, Grain growth and Hall-Petch relation in dual-sized ferrite/cementite steel with nano-sized cementite particles in a heterogeneous and dense distribution. Scripta Mater. **54**(6), 1193–1197 (2006)
5. M. Zhao, T. Hanamura, H. Qui, K. Nagai, K. Yang, Dependence of strength and strength-elongation balance on the volume fraction of cementite particles in ultrafine grained ferrite/cementite steels. Scripta Mater. **54**(7), 1385–1389 (2006)
6. T. Hanamura, F. Yin, K. Nagai, Ductile-Brittle transition temperature of ultrafine ferrite/cementite microstructure in a low carbon steel controlled by effective grain size. ISIJ Int. **44**, 610–617 (2004)
7. M. Zhao, T. Hanamura, H. Qui, H. Dong, K. Yang, K. Nagai, Low absorbed energy ductile dimple fracture in lower shelf region in an ultrafine grained ferrite/cementite steel. Metall. Mater. Trans. A **37A**(9), 2897–2900 (2006)
8. F.B. Pickering, T. Gladman, Metallurgical developments in carbon steels. ISI Spec. Rep. **81**, 10 (1963)

Chapter 2
Ultra-Fine Grained Steel: Relationship Between Grain Size and Tensile Properties

This chapter examines and discusses the characteristics of ultra-fine-grained steel in terms of strength (hardness or tensile yield strength), particularly from the view point of the Hall-Petch relationship, which is an empirical equation that expresses the relationship between the strength and microstructure of steel in terms of an equation in which the yield strength is proportional to the root square of the average grain size in the material examined [1, 2].

2.1 Grain Growth and Hall-Petch Relation in Ultra-Fine-Grained Steel

When certain annealing conditions (such as annealing below the austenitization temperature) are applied to sub-micron-grained ferrite/cementite steel, characteristic grain growth is observed to occur in relation to the grain growth of cementite particles via the Ostwald ripening mechanism. The predicted applicability of the Hall-Petch relation to the hardness and average ferrite size is demonstrated independent of the distribution and size of the cementite particles.

Ultra-fine-grained ferrite/cementite steels have been successfully developed in recent years on the laboratory scale as well as on a pilot scale [3, 4]. In most cases, the microstructure of these low-carbon steels has been generally characterized as the ultra-fine-grained ferrite matrix with nanosized globular cementite particles, where the cementite particles are distributed either by a homogeneous dispersion within the matrix or by a heterogeneous distribution with a local high density.

The Hall-Petch relation, which has been actively examined worldwide, has lately received considerable attention, particularly with regard to ultra-fine-grained materials of several types, including steel, aluminium and titanium [5–11]. Other than the typical/modified Hall-Petch relationship [5–8], the negative slope of this relationship in which the strength/hardness decreases with the grain size has also been reported [9–11]. However, much systematic investigation has not been performed on the Hall-Petch relationship particularly regarding the dependence of the strength/hardness on the ferrite grain size in the ultra-fine-grained ferrite/cementite

© National Institute for Materials Science, Japan. Published by Springer Japan 2014   9
T. Hanamura and H. Qiu, *Analysis of Fracture Toughness Mechanism in Ultra-fine-grained Steels*, NIMS Monographs, DOI 10.1007/978-4-431-54499-9_2

steel with heterogeneously distributed cementite particles (hereafter termed as UFGF/CH steel).

This study attempts to investigate the validity of the Hall-Petch relation in the UFGF/CH steel containing a wide variation of ferrite grain sizes prepared through annealing below the austenitization temperature. The temperature below austenitization was employed because at higher temperatures, transformation to austenite is expected, leading to the annihilation of the ultra-fine-grained ferrite microstructures making structure control impossible. Attempts have also been made to provide a reasonable explanation to the characteristic grain growth that occurs upon annealing in terms of the cementite particle spacing obtained from actual measurements.

The chemical composition (wt%) of the test steel used in this study was determined to be 0.154C, 0.301Si, 1.504Mn, 0.011Al, 0.002S, and 0.001P. By means of the caliber warm-rolling process, some submicron-grained ferrite/cementite microstructures were obtained. Using a vacuum induction furnace, steel ingots were melted in vacuum on a laboratory scale, homogenized for 60 min at 1,473 K, and then hot-forged to rods, 115 mm in diameter. The forged rods were reheated for 60 min at 1,173 K, caliber groove-rolled first at 1,073 K, then at 723 K into 17-mm sized square rods by an accumulated area reduction of 95 % (severe plastic deformation) and a total equivalent strain of nearly 3.0; the rods were then water-cooled. The specimens were cut from the as-rolled rods, isothermally annealed in an argon atmosphere, and water-cooled (Table 2.1) to obtain a wide range of average ferrite-grain-sized parameters. The microstructures of the as-rolled and annealed samples were examined in the transverse section using scanning electron microscopy (SEM) after mechanical grinding, polishing, and etching in a 2 % Nital solution. The average ferrite grain sizes were measured by the linear intercept method. The room temperature measurements of hardness (HV) were conducted by the Vickers hardness test, with an applied load of 100 N.

A typical SEM micrograph of the as-rolled sample is presented in Fig. 2.1a. The microstructure consists of an ultra-fine-grained ferrite matrix and nano-sized (smaller than 100 nm in diameter) globular cementite particles. It was observed that the cementite particles were heterogeneously distributed with a local high density. The representative SEM micrographs of the annealed samples are presented in Fig. 2.1b–f, revealing the growth of ferrite grains as well as the change in the distribution and size of cementite particles upon annealing. The ferrite grain sizes and the distributions and sizes of cementite particles showed almost no change (Fig. 2.1b) at the beginning of annealing at temperatures lower than 773 K, at 773 K

Table 2.1 Annealing procedures for the test steel

Annealing temperature (K)	Annealing time (min)
693	60, 600, 3,000, 6,000
743	60
773	10, 30, 60, 6,000
873	10, 30, 60, 300, 600, 1,200, 3,000, 6,000

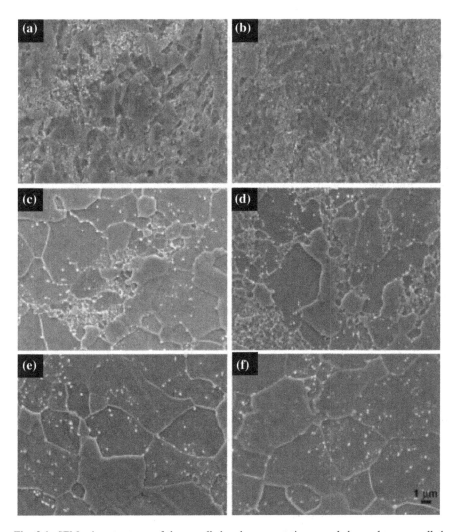

Fig. 2.1 SEM microstructures of the as-rolled and representative annealed samples: **a** as-rolled sample **b** (773 K-60 min) annealed sample **c** (773 K-6,000 min) annealed sample **d** (873 K-30 min) annealed sample **e** (873 K-600 min) annealed sample **f** (873 K-6,000 min) annealed sample (after [1])

for 10, 30 and 60 min, and at 873 K for 10 min. The ferrite grains subsequently began to grow distinctively at 773 K for 6,000 min (Fig. 2.1c), and at 873 K for 30 min or longer times (Fig. 2.1d–f). It is quite interesting to note that although most ferrite particles grew considerably in size at 873 K for 30 min or longer, the microstructures of those in the region with a high density of cementite particles did not change much up to 300 min.

Thereafter, these ferrite grains started coarsening with increase in the annealing time. Under these annealing conditions, some cementite particles coarsened, and the cementite particle spacing increased at the expense of the other cementite particles. By annealing at 873 K for more than 300 min, the area with a local high density of cementite particles eventually disappeared, and the largest cementite particles grew to almost 300 nm. This selective growth of cementite particles is generally referred to as the Ostwald ripening, and is well documented in some references [12–15].

In relation to the above-mentioned characteristic grain growth, the average grain size (d) was not measured by the customary method (summation of the grain sizes divided by the number of grains) in this study, but was measured in terms of $f \times d^{-1}$, i.e., d was determined by

$$d_g^{-1} = f \times d_g^{-1} + (1-f) \times d_{ng}^{-1}, \tag{2.1}$$

where f and d_g are the area fraction and the average ferrite size of steel with the high density of cementite particles, respectively, and d_{ng} is the average ferrite size of steel with the normal density of cementite particles. Figure 2.2 shows the measured values of f, d_g, and d_{ng}, and the changes in d and HV with change in annealing conditions. A curve between the HV and negative square root of d, i.e., HV versus $d^{-1/2}$ is shown in Fig. 2.3, and exhibits a rather monotonic dependence.

The average ferrite size increases with increase in the annealing temperature or time, and accordingly, the hardness decreases with increasing annealing temperature or time, i.e., with increase in the average ferrite grain size. It is also interesting to note that the HV versus $d^{-1/2}$ curve synchronizes well, independent of the distribution and size of the cementite particles. This fact indicates that the hardness of the UFGF/CH steel is in fact dependent on the ferrite grain size, expressed by a single line, thus following a Hall-Petch relation distinctively. A regression equation describing the relationship between the hardness and the average ferrite grain size was developed using all the experimental data and is given as Eq. (2.2) (with the units indicated within the brackets) with a correlation coefficient of 0.99.

$$\mathrm{HV} = 56.55[\mathrm{HV}] + 214.19[\mathrm{HV} \cdot \mu\mathrm{m}^{1/2}]d^{-1/2}[\mu\mathrm{m}^{1/2}] \tag{2.2}$$

From the originally expressed dependence of the yield stress on the grain size, the Hall-Petch relation has been successfully applied to describe the dependence of hardness on the grain size in this ferrite/cementite steel, revealing a typical Hall-Petch relation, and the hardness increased with decrease in the grain size. This implies a potential for large hardening by grain refining for steel types with spherodized cementite particles. As has been described earlier, one of the characteristic features of the microstructure of steel developed in this study is the heterogeneous distribution of its grains, i.e., the nano-sized cementite particles are heterogeneously distributed with a local high density; this condition is believed to accentuate the characteristic grain growth upon annealing.

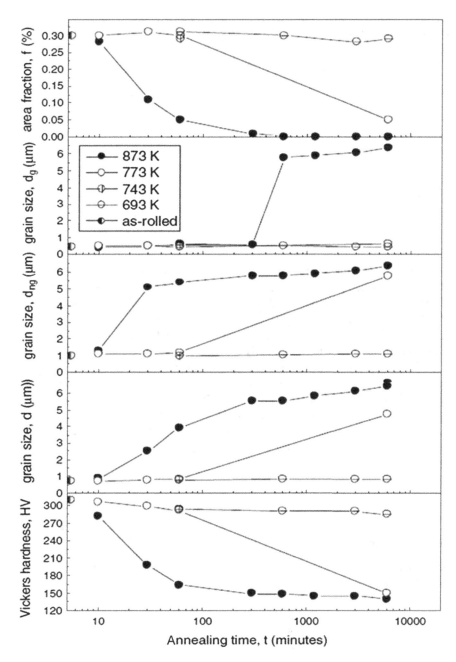

Fig. 2.2 Measured values of f, dg, and dng, and the change in d and HV under various annealing conditions. f is the measured area fraction with a high density of cementite particles, dg is the measured average ferrite size with a high density of cementite particles, dng is the measured average ferrite size with a normal density of cementite particles, d is the average ferrite size, and HV is the Vickers hardness (after [1])

Fig. 2.3 Relation between
Vickers hardness (HV) and
average ferrite size (d) (after
[1])

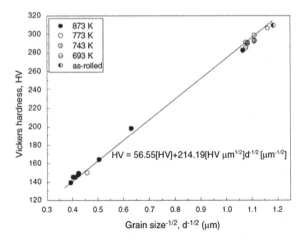

Such a growth process of ferrite grains could be conveniently expressed in terms
of the Larsen-Miller parameter (*P*), describing the combined effects of the annealing
temperature (*T* in Kelvin) and time (*t* in hours), which is defined as
$P = T (20 + 1 \, gt)$ [14, 15]. The HV as a function of *P* is shown in Fig. 2.4, which
depicts the dependence of hardness on the combined effects of the annealing
temperature and time. The HV abruptly decreased when *P* becomes approximately
17×10^3 (e.g., annealing at 873 K for 30 min), and settled at an almost constant
value of HV140. The elongated ferritic structure due to heavy deformation was
fully retained (Fig. 2.1a and b) before the softening began. However, upon softening some grains with the elongated structure disappeared and equiaxed ferrite
grains began to form (Fig. 2.1c–f). The ferrite grains tended to grow gradually when
the hardness is decreased (Figs. 2.2 and 2.3).

Fig. 2.4 Change in Vickers
hardness with annealing as a
function of the annealing
parameter (P); T and t denote
Kelvin and hours,
respectively (after [1])

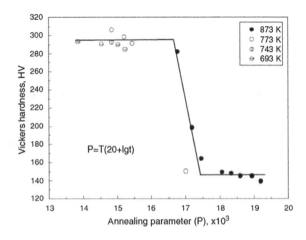

Fig. 2.5 Cementite particle spacing and average ferrite size measured at 873 K with the changed annealing time for the sites with a high density of cementite particles (after [1])

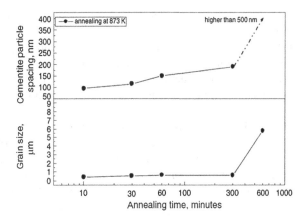

The coarsening of the ferrite grains in the region with a high density of cementite particles could be explained as follows. In this region, the ferrite grain boundaries are pinned by the nano-sized cementite particles and are forced to bulge when they grow in size. When the cementite particles densely cover the ferrite grain boundary, the grain boundary migration can be effectively obstructed, thereby forcing the coarsening of ferrite grains to slow down. Along with the ferrite particle coarsening, the distribution and size of the cementite particles change upon annealing, as illustrated in Fig. 2.1.

This microstructure change would eventually interact with ferrite coarsening. Figure 2.5 shows the measured cementite particle spacing and the average ferrite grain size for the sites with high-density cementite particles at 873 K for varied annealing times. At the beginning of annealing, when the temperature was lower than 873 K or at 873 K for 10 min, cementite particle spacing as well as their corresponding ferrite grain sizes did not change remarkably. The Ostwald ripening occurred subsequently when annealed at 873 K by maintaining the temperature constant until 300 min, and the cementite particle spacing increased, while the ferrite grain sizes with the highly dense cementite particles hardly changed. This indicates that the cementite particle spacing was not sufficiently high to bulge the ferrite grain boundaries. Thereafter, by annealing at 873 K and by holding this temperature for 600 min, the local high density region of cementite particles disappeared and the cementite particle spacing increased distinctively because of the Ostwald ripening of the cementite particles and the high growth observed in the size of the ferrite particles. This indicates that the cementite particle spacing is closely related to the ferrite grain boundary bulge. Hence, ferrite grain coarsening does not occur until cementite particle spacing reaches some certain potential critical value, which is involved in the Ostwald ripening of the cementite particles upon annealing.

Thus, the following conclusions can be drawn from the discussion so far in this chapter.

(1) The characteristic ferrite grain growth occurs along with the Ostwald ripening of cementite particles upon annealing in submicron-grained ferrite/cementite steel with a heterogeneous and dense distribution of cementite particles. This growth behavior can be explained in terms of the cementite particle spacing obtained from the actual measurements. The ferrite coarsening does not occur until the cementite particle spacing reaches some potential critical value.

(2) Despite the fact that the hardness as well as the average ferrite grain size is influenced by the distribution and size of cementite particles, a plot between the hardness and the negative square root of the average ferrite grain size synchronizes well, and is independent of the distribution and size of the cementite particles. The predictive capability of the Hall-Petch relation between the hardness and the average ferrite grain size has been demonstrated, implying that a significant potential for hardening exists by grain refinement.

2.2 Dependence of the Strength and Strength-Elongation Balance on Volume Fraction of Cementite Particles in Ultrafine Grained Ferrite/Cementite Steels

An increase in the cementite volume fraction increases the athermal component of strength. The strain-hardening rate increases with increase in the carbon content up to 0.3 wt% C and then changes very little with a further increase in the carbon content, which is well reflected in the change of the uniform elongation with carbon content. The strength-elongation balance could be practically improved by means of an appropriate cementite volume fraction.

Ultra-fine-grained C–Mn steels have great commercial potential as an advanced structural material because of their high tensile strength and low ductile-to-brittle transition temperature, in addition to the environmental advantage they do not require any expensive alloying elements.

So far, ultra-fine-grained C–Mn steels with an average ferrite grain size of smaller than a few micro meters and even smaller than 1 μm have been successfully developed in the laboratory and/or on a pilot scale, and are generally characterized by an ultra-fine-grained ferrite matrix and finely globular cementite particles (hereafter termed as UGF/C steels) [1, 3, 4]. The volume fraction of the cementite particles, which is closely related to the carbon content of the steel, is undoubtedly one of the critical factors affecting the tensile properties of the UGF/C steels with spherodized cementite particles.

Structural steels must meet the industrial requirement in terms of the tensile properties. The gain size necessary to achieve a certain value of tensile strength has been realized in ultrafine-grained steels. However, it is still discouraging that most of these steels exhibit a total elongation of less than a few percent in tensile tests, sometimes giving a uniform elongation of almost zero percent [4–6]. Such deficiency in mechanical properties is one of the important factors that have restricted

Table 2.2 Measured chemical composition of the test steels (wt%) (after [2])

Steel (wt % C)	C	Si	Mn	S	P	N	O	Cu	Al
0.05	0.054	0.310	1.492	0.002	0.001	0.0013	0.0036	<0.001	<0.001
0.15	0.148	0.311	1.506	0.002	0.001	0.0012	0.0014	<0.001	0.010
0.3	0.299	0.311	1.500	0.002	0.001	0.0011	0.0011	<0.001	0.013
0.45	0.448	0.313	1.509	0.002	0.001	0.0012	0.0009	<0.001	0.016
0.6	0.600	0.311	1.511	0.002	0.001	0.0010	0.0008	<0.001	0.017

the potential commercialization of UGF/C steels. The improvement in the strength-elongation balance is considered to be one of the current major concerns in the UGF/C steels technologies. This section examines the effect of the volume fraction of cementite particles on both the tensile strength and strength-elongation balance in UGF/C steels.

2.2.1 Experimental

The chemical compositions (in wt%) of the test steels are listed in Table 2.2. The difference in the chemical compositions among these steels is mostly only in the carbon content (varying from 0.05 to 0.6 wt%), with little difference with respect to the other alloying elements. The UGF/C microstructures were obtained by caliber-warm-rolling. The ingots were melted in vacuum on a laboratory scale, homogenized for 60 min at 1,473 K, hot-forged to rods 115 mm in diameter, and then subjected to multi-pass caliber-warm-rolling to obtain rods 17 mm in diameter by an accumulated area reduction of approximately 95 % (severe plastic deformation). The outline of this production process is illustrated in Fig. 2.6.

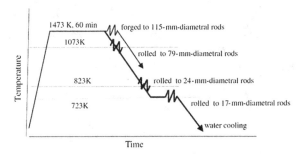

Fig. 2.6 Outline of the process. The samples were homogenized for 60 min at 1,473 K, hot-forged to 115-mm-diametral rods, and then subjected to multipass caliber-warm-rolling to the final 17-mm-diametral rods by an accumulated area reduction of approximately 95 % (severe plastic deformation) (after [2])

Cylindrical tensile test pieces, whose lengths were equal to the rod length, were machined from the rods with a gauge diameter of 3.5 mm and a gauge length of 25 mm. Tensile tests were conducted with a cross-head speed of 0.5 mm/min at different temperatures: 323 K, room temperature (293 K), 210 K, and 77 K. Their microstructures were observed in the transverse directions using an SEM.

The SEM images are presented in Fig. 2.7: the UGF/C microstructures (Fig. 2.7a–e) and the change in the volume fraction of the cementite particles (Fig. 2.7f) with various carbon contents. As is shown in Fig. 2.7a–e, the ultra-fine-grained ferrite matrices were populated by numerous nano-sized globular cementite particles. The ferrite grains are slightly elongated in the rolling direction because of the applied heavy deformation. The average ferrite grain sizes changed insignificantly on varying the carbon content, from 0.6 to 0.05 wt% C. The cementite particles with diameters of much less than 100 nm were distributed somewhat heterogeneously together with a local high density.

The characteristic distribution of the cementite particles could be attributed primarily to the original microstructure, because the caliber-warm-rolling process starts from the ferrite and pearlite microstructures, and the cementite particles naturally localize within the original region of the pearlite structures. The volume fraction of the cementite particles depends on the carbon content. The higher the carbon content, the larger is the volume fraction of the cementite particles, as illustrated in Fig. 2.7a–e. The volume fractions of the cementite particles for the test steels with various carbon contents were roughly evaluated by thermodynamic equilibrium calculations as shown in Fig. 2.7f. Meanwhile, the distribution of the cementite particles became relatively homogenized and disordered in accordance with a gradual increase in the carbon content from 0.05 to 0.6 wt%.

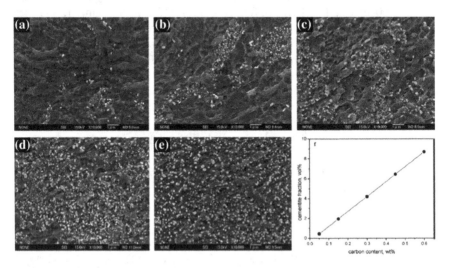

Fig. 2.7 SEM images of UGF/C microstructures and changes in the volume fraction of the cementite particles with various carbon contents (after [2]). **a** 0.05 % C, **b** 0.15 % C, **c** 0.3 % C, **d** 0.45 % C, **e** 0.6 % C

Fig. 2.8 Nominal stress-strain curves after the tensile test (after [2])

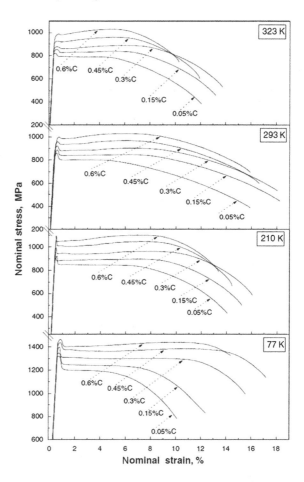

The nominal stress-strain curves after the tensile test are presented in Fig. 2.8, which depicted discontinuous yielding followed by a prolonged small strain hardening with a considerable uniform elongation. At each tensile temperature, the lower yield stress (LYS), upper yield stress (UYS), and ultimate tensile stress (UTS) clearly increased with the carbon content (wt% C), i.e., the volume fraction of the cementite particles.

The LYS values obtained from the nominal stress-strain curves as illustrated in Fig. 2.8, are plotted as a function of wt% C at each tensile temperature (see Fig. 2.9a). The LYS versus wt% C curve synchronizes well at each tensile temperature, being expressed by a single line. This clearly indicates that the LYS is linearly dependent on the wt% C. The regression equations describing the relationship between LYS and wt% C are as follows:

Fig. 2.9 a LYS/UYS/UTS change with carbon content at every tensile temperature (after [2]). **b** LYS/UYS/UTS change with tensile temperature at every carbon content (after [2])

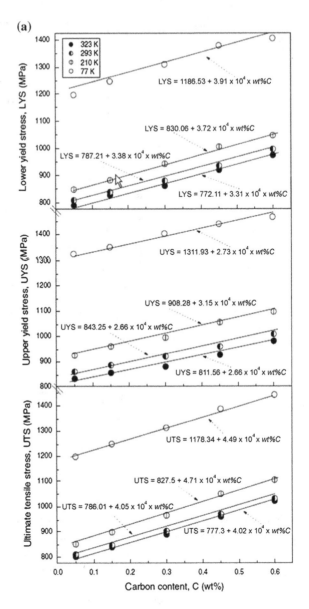

$$LYS = 772.11 + 3.31 \times 10^4 \times wt\% \, C \qquad (2.3)$$

with a correlation coefficient of 0.996 at 323 K,

$$LYS = 787.21 + 3.38 \times 10^4 \times wt\% \, C \qquad (2.4)$$

Fig. 2.9 (continued)   **(b)**

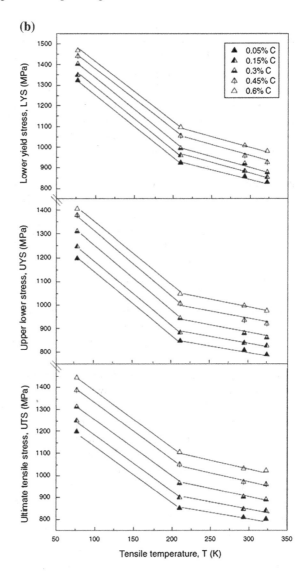

with a correlation coefficient of 0.996 at room temperature,

$$LYS = 830.06 + 3.72 \times 10^4 \times wt\% \, C \qquad (2.5)$$

with a correlation coefficient of 0.997 at 210 K and

$$LYS = 1186.53 + 3.91 \times 10^4 \times wt\% \, C \qquad (2.6)$$

with a correlation coefficient of 0.986 at 77 K.

Because the lines corresponding to the LYS versus wt% C relationship at individual tensile temperatures were parallel to each other as shown in Fig. 2.9a, the difference in the LYS among the LYS versus wt% C curves did not change much with the changes in the tensile temperature. For example, by increasing the carbon content from 0.05 to 0.6 wt%, the LYS increased from 789.8 to 974.5 MPa at 323 K (by 184.7 MPa), from 808.8 to 996.1 MPa at room temperature (by 187.3 MPa), from 848.1 to 1,047.1 MPa at 210 K (by 199 MPa), and from 1,196.7 to 1,405.8 MPa at 77 K (by 209.1 MPa).

The effect of the volume fraction of cementite particles, i.e. the effect of the carbon content, on the LYS is, therefore, only dependent on the volume fraction of the cementite particles or the carbon content itself, and not on the temperature.

The yield strength of materials is known to consist of two components: thermal stress and athermal stress [20]. The effect of the volume fraction of cementite particles, i.e., the effect of the carbon content, accordingly, contributed only to the athermal component of the LYS, and the increase in the volume fraction of the cementite particles or the carbon content increased the athermal component of the LYS. Presuming the looping of dislocation from the cementite particles described by Orowan [21], a higher stress is needed for steels with a higher volume fraction of cementite particles. This means that a higher volume fraction of cementite particles results in a higher athermal component of the LYS. The thermal component of the LYS can be discussed by the difference in the LYS obtained from the LYS versus T curves.

The change in the LYS with the tensile temperature (T) is presented for cases with different carbon contents, as shown in Fig. 2.9b. Because these LYS versus T curves were parallel to each other, the difference in the LYS obtained from the LYS versus T curves did not change significantly by the change in the carbon content, i.e., the volume fraction of the cementite particles. For example, on decreasing the tensile temperature from 323 to 77 K, the LYS increases from 789.8 to 1,196.7 MPa at 0.05 wt% C (by 406.9 MPa), from 826.9 to 1,247.1 MPa at 0.15 wt% C (by 420.2 MPa), from 861.8 to 1,309.6 MPa at 0.3 wt% C (by 447.8 MPa), from 919.9 to 1,378.8 MPa at 0.45 wt% C (by 458.9 MPa), and from 974.5 to 1,405.8 MPa at 0.6 wt% C (by 431.3 MPa). An increase in the volume fraction of the cementite particles, accordingly, contributes to an increase in the athermal component of the LYS, without changing the thermal component of the LYS.

The values for the UYS and UTS are also plotted as a function of wt% C at each tensile temperature in Fig. 2.9a. The lines corresponding to the UYS versus wt% C relationships or the UTS versus wt% C relationships at the various tensile temperatures are parallel to each other. The changes in the UYS and UTS with the tensile temperature (T) are also illustrated for different carbon contents in Fig. 2.9b. The UYS versus T curves or the UTS versus T curves are parallel to each other.

Therefore, an increase in the volume fraction of the cementite particles contributed to an increase in the athermal component of the UYS and UTS, without changing their thermal component, which is similar to the above mentioned effect

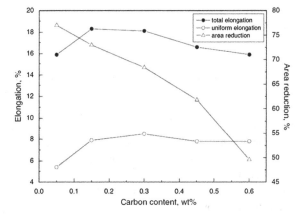

Fig. 2.10 Total elongations, uniform elongations, and area reductions after the room-temperature tensile test (after [5])

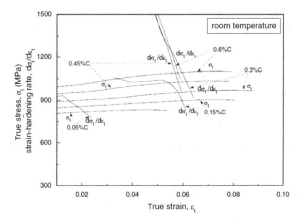

Fig. 2.11 Calculated true stress-strain curves before necking as well as work-hardening rate as a function of true strain (after [2])

of the volume fraction of the cementite particles on the LYS. The total elongations, uniform elongations, and area reductions after the room temperature tensile test are illustrated in Fig. 2.10. The total elongations and uniform elongations increased with the carbon content to up to 0.3 wt% and then began to slightly decrease.

However, the area reduction clearly decreased with increase in the carbon content. The calculated true stress-strain curves before the occurrence of necking as well as the strain-hardening rate as a function of the true strain are shown in Fig. 2.11, which were obtained from identical room temperature tensile tests shown in Fig. 2.8. The true stress seemed apparently extremely sensitive to the carbon content, i.e., the volume fraction of the cementite particles.

An increase in the carbon content, i.e., an increase in the volume fraction of cementite particles, increased the true stress, as shown in Fig. 2.11. However, strain hardening increased on increasing carbon content to up to 0.3 wt% C, after which it remained almost constant on further increasing the carbon content. For a larger uniform elongation, strain hardening was required, because the onset of plastic instability in tension, i.e., the condition of necking in tension, is governed by the following formula:

$$\sigma \geq d\sigma/d\varepsilon, \qquad (2.7)$$

where σ is the flow stress, ε is the true strain, and $d\sigma/d\varepsilon$ is the strain-hardening rate [20–24]. Therefore, the uniform elongation is controlled not only by the flow stress but also by the strain-hardening rate; accordingly, a larger strain-hardening rate is beneficial for improving the uniform elongation. As shown in Fig. 2.11, the increase in the carbon content from 0.05 wt% C, 0.15 wt% C and then to 0.3 wt% C not only increased the flow stress but also the strain-hardening rate, thus improving the strength-uniform elongation balance for the UGF/C steels. Subsequently, a further increase in the carbon content from 0.3, 0.45, and up to 0.6 wt% C definitely increased the flow stress, without significantly changing the strain-hardening rate, which slightly decreased the uniform elongation.

However, the area reduction decreased with increase in the carbon content, as shown in Fig. 2.10. It is interesting to note that the UGF/C steel with 0.05 wt% C did not exhibit good uniform elongation, because the volume fraction of the cementite particles was excessively small to exhibit a high strain-hardening rate. UGF/C steels with a higher carbon content of 0.15 or 0.3 wt% C showed a superior balance of strength and uniform elongation and a small loss of area reduction to approximately 70 %, which is hopefully acceptable in industrial applications.

Based on the above discussions, we can draw the following conclusions:

1. The LYS, UYS, and UTS have been revealed to have monotonic relationships with the carbon content, i.e., the volume fraction of the cementite particles. More specifically, an increase in the volume fraction of the cementite particles increased in the athermal component of these strength parameters, without causing much change in their thermal component.
2. The true stress increased with increase in the carbon content, while the strain-hardening rate increased with increase in the carbon content to up to 0.3 wt% C only and then remained almost constant with further increase in the carbon content, which correlated with a similar change in the uniform elongation. UGF/C steels with a higher carbon content of 0.15 or 0.3 wt% C exhibit a superior balance of strength, uniform elongation, and a small loss of area reduction to approximately 70 %, which may be acceptable for industrial applications.

References

1. M. Zhao, T. Hanamura, H. Qui, K. Nagai, K. Yang, Grain growth and Hall-Petch relation in dual-sized ferrite/cementite steel with nano-sized cementite particles in a heterogeneous and dense distribution. Scripta Mater. **54**(6), 1193–1197 (2006)
2. M. Zhao, T. Hanamura, H. Qui, K. Nagai, K. Yang, Dependence of strength and strength-elongation balance on the volume fraction of cementite particles in ultrafine grained ferrite/cementite steels. Scripta Mater. **54**(7), 1385–1389 (2006)

3. R. Song, D. Ponge, R. Kaspar, The microstructure and mechanical properties of ultrafine grainedplain C-Mn steels. Steel Res. Int. **75**, 33–37 (2004)
4. K. Nagai, Ultrafine-grained ferrite steel with dispersed cementite particles. J. Mater. Process. Technol. **117**(3), 329–332 (2001)
5. N. Hansen, Hall-Petch relation and boundary strengthening. Scripta Mater. **51**(8), 801–806 (2004)
6. N. Tsuji, Y. Ito, Y. Saito, Y. Minamino, Strength and ductility of ultrafine grained aluminum and iron produced by ARB and annealing. Scripta Mater. **47**(12), 893–899 (2002)
7. R.Z. Valiev, F. Chmelik, F. Bordeaux, G. Kapelski, B. Baudelet, The Hall-Petch relation in submicro-grained Al-1.5 % Mg alloy. Scr. Metall. Mater. **27**(7), 855–860 (1992)
8. G.W. Nieman, J.R. Weertman, R.W. Siegel, Tensile strength and creep properties of nanocrystalline palladium. Scripta Metall. Mater. **24**(1), 145–150 (1990
9. K. Lu, W.D. Wei, J.T. Wang, Microhardness and fracture properties of nanocrystalline Ni, P alloy, Scripta Metall. Mater. **24**(12), 2319–2323 (1990)
10. J.S.C. Jang, C.C. Koch, Melting and possible amorphization of Sn and Pb in Ge/Sn and Ge/Pb mechanically milled powders. J. Mater. Res. **5**(2), 325–333 (1990)
11. A.H. Chokshi, A. Rosen, J. Karch, H. Gleiter, On the validity of the hall-petch relationship in nanocrystalline materials. Scr. Metall. **23**(10), 1679–1683 (1989)
12. K.M. Vedula, R.W. Heckel, Spheroidization of binary Fe-C alloys over a range of temperatures. Metall. Trans. **1**(1), 9–18 (1970)
13. GR Speich, Tempering of low-carbon martensite. Trans TMS-AIME **245**, 2553–2564 (1969)
14. G.P. Airey, T.A. Hughes, R.F. Mehl, The growth of cementite particles in ferrite. Trans TMS-AIME **242**, 1853–1863 (1968)
15. E. Nes, WB Hutchinson, in *Bilde-Sorensen Proc of the 10th Riso Symposium*, Riso National Laboratory, Roskilde, Denmark, (1989) p. 233
16. J.H. Bucher, J.D. Grozier, J Iron Steel Inst **204**, 1253 (1966)
17. R.A. Grange, R.W. Baughman, Trans. Am. Soc. Metall. **48**, 165 (1956)
18. N. Matsukura, S. Nanba, in Proceedings of the first symposium on super metal, Tokyo, Japan, November, 1998, R&D Institute of Metals and Composites for Future Industries and Japan Research and Development Center for Metals, (1998) p. 229
19. C.C. Koch, D.G. Morris, K. Lu, A. Inoue, Ductility of nanostructured materials. MRS Bull. **24** (2), 54–58 (1999)
20. T. Tsuchiyama, H. Uchida, K. Kataoka, S. Takaki, Fabrication of fine-grained high nitrogen austenitic steels through mechanical alloying treatment. ISIJ Int. **42**(12), 1438–1443 (2002)
21. E. Orowan, *Internal Stress in Metals and Alloys* (Institute of Metals, London, 1948), p. 451
22. E.W. Hart, Theory of the tensile test. Acta Metall. **15**(2), 351–355 (1967)
23. G.E. Dieter, *The Plastic Deformation of Metals, Mechanical metallurgical*, 3rd edn. (McGraw-Hill, Boston (MA), 1986), p. 290
24. D. Jia, Y.M. Wang, K.T. Ramesh, E. Ma, Y.T. Zhu, R.Z. Valiev, Deformation behavior and plastic instabilities of ultrafine-grained titanium. Appl. Phys. Lett. **79**, 611–613 (2001)

Chapter 3
Ultra-Fine-Grained Steel: Relationship Between Grain Size and Impact Properties

This chapter presents the study on the relationship between the effective grain size (d_{EFF}) and the ductile-to-brittle transition temperature (DBTT) in impact tests. In addition, the low absorbed energy with a ductile dimple fracture in the lower shelf region, a characteristic feature of the ultra-fine- grained ferrite/cementite (UGF/C) steel is presented. In addition, this chapter also shows that a transition from an energy-absorbent ductile mode to an energy-absorbent brittle mode in impact tests exists [1, 2].

3.1 DBTT Controlled by Effective Grain Size of Ultra-Fine Ferrite/Cementite (UGF/C) Microstructure in Low Carbon Steels

With the aim of analyzing the characteristic excellent toughness of UGF/C steels, the concept of d_{EFF} is employed and applied to DBTT for UGF/C, ferrite/pearlite (F/P), quenched (Q), and quench-and-tempered (QT) microstructures in low carbon steels. The value of d_{EFF} is determined to be 8, 20, 100, and 25 μm for UGF/C, F/P, Q, and QT, respectively. In F/P and Q, the value was in accordance with the ferrite grain size and the prior austenite grain size, respectively. In QT, the d_{EFF} value fits with the martensite packet size. In UGF/C, however, the ferrite grain size showed a bimodal distribution and the larger grain size corresponded to the d_{EFF} value, which was the smallest among the four microstructures.

In terms of the relationship between d_{EFF} and DBTT, the UGF/C, Q, and QT microstructures can be placed into the same group and the F/P microstructures to a different group. Furthermore, the UGF/C microstructures show the highest estimated fracture stress among the four microstructures. This behavior might be the result of the difference in the surface energy of fracture, the UGF/C is estimated to have a surface energy of 34.6 J/m² and the F/P a surface energy of 7.7 J/m². Thus, the excellent toughness of the UGF/C steel can be attributed to the small d_{EFF} and high surface energy of fracture.

© National Institute for Materials Science, Japan. Published by Springer Japan 2014   27
T. Hanamura and H. Qiu, *Analysis of Fracture Toughness Mechanism in Ultra-fine-grained Steels*, NIMS Monographs, DOI 10.1007/978-4-431-54499-9_3

Fig. 3.1 Relationship
between change in DBTT and
change of 9.8 MPa in yield
strength for different
microstructural factors, i.e.,
dislocation strengthening,
dispersion strengthening, and
grain-size strengthening (after
[6])

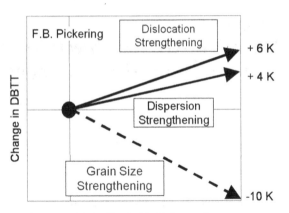

The improvement in the trade-off balance between strength and toughness has
been a major concern for structural materials. The balance between yield strength
(YS) and DBTT has so far attracted a great deal of attention from an engineering
point of view. In this aspect, Pickering [3] derived the balance vectors with respect
to the strengthening mechanism from his collected database for low-carbon steels
with ferrite grain sizes larger than 10 μm and pointed out the significant performance caused by grain refinement. As can be seen in Fig. 3.1, the change in DBTT
by an increase of 9.8 MPa in the YS is +6 K in dislocation strengthening, +4 K in
dispersion strengthening, and +10 K in grain-size strengthening. In the F/P of low-
carbon steels, DBTT can be determined using the crystallographic ferrite grain size,
d, and expressed in the following forms [3–6]:

$$\text{DBTT} = A - K d^{-1/2} \text{ or} \tag{3.1}$$

$$= A - K \ln d^{-1/2} \tag{3.2}$$

where the constant A contains metallurgical factors other than the grain size and the
coefficient K is independent of the grain size. The UGF/C steel shows an excellent
balance of high YS and low DBTT as expected. However, the above-mentioned
equation is not necessarily applicable [7] when the ferrite grain size is taken as d. In
other words, the value of DBTT varies widely between Eqs. (3.1) and (3.2) when
the grain size is in the range of a few μm or smaller. The DBTT estimated with
$d = 1$ μm from Eq. (3.1) is 49 K, while that from Eq. (3.2), based on the extrap-
olation of the experimental data from reference [8], is 104 K.

The 9 % Ni steel is an excellent material with a good balance of strength and
toughness [9]. The microstructure of this steel is similar to that of QT martensite,
though it occasionally contains a small volume of the retained austenite. Because Ni
addition does not increase the YS by itself, its high YS is believed to be inherent to
the QT martensite. Hence, the specific effect of Ni is believed to enhance the

toughness of QT steel, whose mechanism of is still disputable. In the QT steel, the idea of d_{EFF} [10, 11] accounts for the DBTT in the steel (Terasaki et al. used the term "unit crack path" [12], instead of "effective grain size").

According to the idea of Griffith [20], the fracture stress (σ_F) is a function of the surface energy of fracture and an imaginary crack size. The unstable growth of the Griffith crack has been discussed by assuming that the imaginary pre-existing crack interacting with the stress field in the transition temperature range is comparable to the size of the ferrite grain [13]. Under the same stress field in the transition temperature range, plastic deformation prevails over brittle fracture when yielding occurs first or the brittle fracture dominates when the fracture starts before yielding. Because one criterion that determines the initiation of yielding in a polycrystalline material is the grain size or slip length, it is simple and reasonable to suppose that the cleavage cracking of a grain is a counter measure to initiate cleavage fracture. However, such a "grain" idea is not always evident in all steel microstructures.

The brittle fracture belongs to the family similar to transgranular cleavage in body centered cubic (bcc) steels. In the F steel, the brittle fracture is a typical cleavage of the $\{100\}_\alpha$ plane [14]. Furthermore, the cleavage planes have also been identified as $\{100\}_\alpha$ [15–17] in F/P, QT, and UGF/C steels. Hence, we can finally understand that d_{EFF}, in which a cleavage crack moves through in a straight manner, corresponds to the microstructural unit having a specific crystallographic orientation and bounded by high-angle boundaries. In addition, d_{EFF} can be determined experimentally by fractography as well.

Flow stress increases at lower temperatures. The fracture stress (σ_F) which is assumed to be equal to the flow stress at DBTT, is believed to be independent of temperature. Hence, an increase in flow stress or YS increases the temperature at which the flow stress reaches σ_F or shifts the DBTT to a higher temperature. In addition, a further increase in DBTT may occur when the increase in YS decreases σ_F, as described in Fig. 3.2a [3]. Unless the grain refinement increases σ_F significantly, the shifting of DBTT to a lower temperature cannot be accounted for as demonstrated in Fig. 3.2b. Thus, the reasons for UGF/C generating the excellent toughness should also be considered in terms of σ_F.

In the present study, we fabricated four different microstructures from a single steel with low carbon content and extensively examined the determination of DBTT for each microstructure with the aid of d_{EFF}, paying special attention to the excellent toughness of the UGF/C steel. Recently, UGF/C has sometimes been shown to have bimodal structures with two peaks existing in the grain size distribution [17]. Hence, it is also interesting to determine which crystallographic unit controls the DBTT mainly in the UGF/C structure.

With a vacuum induction furnace, an ingot was melted in vacuum in the laboratory scale. The chemical composition analogous to the conventional JIS-SM490 is shown in Table 3.1. The ingot was homogenized for 1,800 s at 1,473 K and hot-rolled to a 23-mm thick plate, and the following samples were fabricated. (1) F/P samples were machined from the plate. (2) UGF/C samples were produced from the plate in the following manner. The plate was heated for 1,800 s at 1,173 K, followed by water-quenching; then, the same samples were caliber warm-rolled into

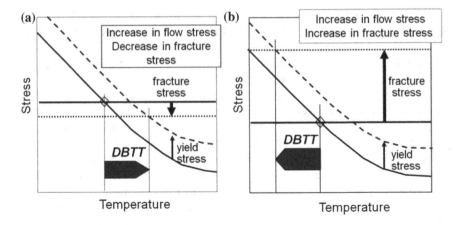

Fig. 3.2 a Schematic relationship among yield stress, fracture stress, and DBTT: in the case where a further increase in DBTT occurs when the increase in YS decreases σ_F. **b** Schematic relationship among yield stress, fracture stress, and DBTT: in the case where a further decrease in DBTT occurs when the increase in YS significantly increases σ_F (after [6])

Table 3.1 Chemical compositions of the steel (after [1]) (mass %)

	C	Si	Mn	P	S	T–Al	T–N
Present	0.148	0.33	1.41	0.001	0.001	0.003	0.0012
SM 490	0.15	0.28	1.45	0.005	0.0004	0.031	0.0016

12-mm squares and 1,000-mm long rods by an accumulated area reduction of 85 %. (3) Q samples, designated as quenched, were the UGF/C samples annealed for 3,600 s at 1,373 K and water quenched. Finally, (4) the QT samples were obtained by tempering the Q samples for 3.65 s at 723 K.

The cylindrical tensile test pieces were machined from these four types of samples with a gage diameter of 3.5 mm and a gage length of 25 mm. The direction of the tensile test pieces was parallel to the rod length. Charpy impact test pieces were similarly prepared in a standard 10-mm square, 55-mm long, and 2-mm V-notched geometry. The length of the Charpy impact test pieces was parallel to the rod length and the notch was vertical to both the rod length and one of the four flanks of the rod. Tensile tests were performed with a crosshead speed of 0.5 mm/min at room temperature, and the Charpy impact tests were carried out at the temperature range 78–373 K.

Samples from the Charpy impact test were observed with an optical microscope (OM), a scanning electron microscope (SEM), by electron backscattered diffraction (EBSD), and with a transmission electron microscope (TEM). The d_{EFF} value was determined by measuring the cleavage fracture unit of each sample fractured at 77 K. Each fractured sample was cut in half, embedded into a plastic mold, and observed by SEM.

3.1.1 Microstructure

The SEM microstructures for all the samples are shown in Fig. 3.3. In the F/P sample, the horizontal and normal directions were parallel and perpendicular to the hot-rolling direction, respectively. In the other samples, the normal direction identical to the caliber-rolling direction or the cross-sectional area is shown. The F/P steel (a) consisted of ferrite and pearlite. White parts corresponded to pearlite and the gray parts to ferrite. The average ferrite grain size was 20 μm. According to the SEM observations, the UGF/C (b) consisted of equiaxed ultrafine grain microstructures and finely dispersed cementites. Here, white dots corresponded to the cementites and the gray parts to the ferrite. The ferrite grains were elongated to some extent in the longitudinal direction. However, the EBSD analysis also characterized the grain boundaries in the UGF/C sample, as shown in Fig. 3.4a. In this figure, the ferrite grains were manifested by the boundaries with the misorientation above 5°. The microstructure had two types of characteristic ferrite grains with peak area fraction: the fine ones (approximately 1.6 μm) and the coarse ones (approximately 8 μm) [18]. The average grain size determined by grain area averaging, including the two types of characteristic grain sizes, was 2.7 μm.

Q (c) in Fig. 3.3 consisted mostly of martensite, while QT (d) in Fig. 3.3 consisted mostly of tempered martensite. The large unit corresponded to the prior austenite grain, which, for both samples, was determined to be approximately 100 μm. In this study, the EBSD analysis could be applied to the Q sample because of the large strains in the matrix that prevented precise EBSD measurements. However, the detailed microstructure of QT showed the interior of the prior austenite grain, where the fine martensite laths and fine cementite particles could be observed. The EBSD analysis also characterized the grain boundaries in the QT sample, and the grains were manifested by the boundaries with misorientation above 5°, as shown in Fig. 3.4b. The microstructure had grains with peak area fractions with a size of approximately 25 μm, which is believed to correspond to the packet size.

3.1.2 Charpy Impact Tests

Table 3.2 lists the YS, TS, total elongation (T.El.), and uniform elongation (U.El.) for each structure. The Charpy impact absorption energy is plotted as a function of test temperature for the four samples in Fig. 3.5. Q showed the poorest impact properties among the samples. In this particular case, the upper shelf energy could not be determined even at 373 K. In contrast, QT showed improved properties of 240 K for the absorbed energy transition temperature and 240 J for the upper-shelf energy. F/P showed a lower absorbed energy transition temperature and a higher upper-shelf energy than QT. UGF/C showed the best impact properties with the

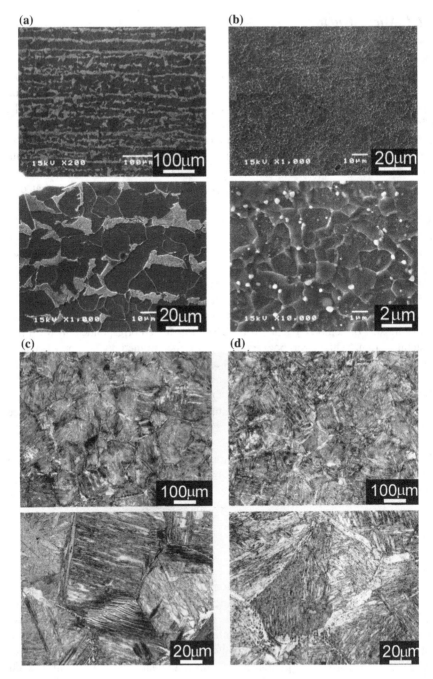

Fig. 3.3 Microstructures of **a** ferrite/pearlite (F/P), **b** ultrafine ferrite/cementite (UGF/C), **c** as-quenched (Q), and **d** quench-and-tempered (QT) steels (after [1])

(a)

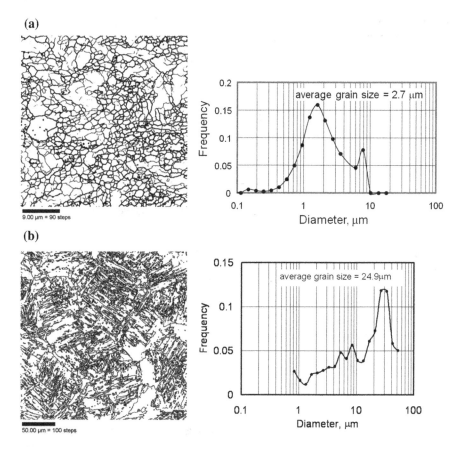

(b)

Fig. 3.4 **a** EBSD analysis on the ultrafine ferrite/cementite (UGF/C) steel. *Right* Ferrite grains manifested by the boundaries with the misorientation above 5° (the difference in crystallographic orientation is 5° and more). *Left* Grain size and misorientation distribution calculated by the analysis (after [1]). **b** EBSD analysis of the quench-and-tempered (QT) steel. *Right* Ferrite grains manifested by the boundaries with the misorientation above 5° (the difference in crystallographic orientation is 5° and more). *Left* Grain size and misorientation distribution calculated by the analysis (after [1])

Table 3.2 Properties of the SM490-equivalent steel with different microstructures, i.e., ferrite/pearlite (F/P), ultra-fine ferrite/cementite (UGF/C), as-quenched (Q), and quench-and-tempered (QT) steels: YS (yield strength), TS (tensile strength), T.El. (total elongation), and U.El. (uniform elongation) (after [1])

Sample	YS (MPa)	TS (MPa)	T.El. (%)	U.El. (%)
F/P	261	463	35	16
UGF/C	536	612	21	9.5
Q	619	1,027	4	2.9
QT	548	676	17	7.4

Fig. 3.5 Charpy impact
absorbed energy as a function
of test temperature for ferrite/
pearlite (F/P), ultra-fine
ferrite/cementite (UGF/C), as-
quenched (Q), and quench-
and-tempered (QT) steels
(after [1])

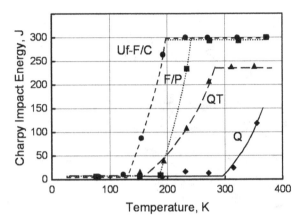

Fig. 3.6 Crystallinity after
Charpy impact test as a
function of test temperature
for ferrite/pearlite (F/P), ultra-
fine ferrite/cementite (UGF/
C), as-quenched (Q), and
quench-and-tempered (QT)
steels (after [1])

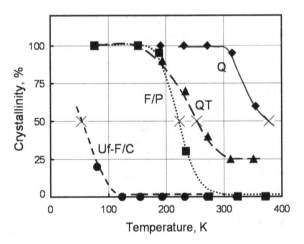

lowest absorbed energy transition temperature of 160 K while maintaining the high
upper-shelf energy level of 300 J at test temperatures down to 230 K.

The change in crystallinity or the area percent of the brittle fracture on the
fracture surface measured by fractography is plotted as a function of test temperature in Fig. 3.6. In this study, DBTT is defined as the temperature with 50 %
crystallinity. The DBTT for the four samples were 53 K for UGF/C, 223 K for F/P,
253 K for QT, and 378 K for Q. Figure 3.7 shows the SEM image of the side view
of the fracture surface for each sample tested at 77 K. In each case, the crack path
was from the right to the left direction, with the V-notch at the right end. Therefore,
the fracture surface was observed as a line at the top-edge side view of the sample.
The value was determined to be 8, 20, 100, and 25 μm for UGF/C, F/P, Q, and QT,
respectively. In the case of F/P and Q, the values were in good agreement with the
ferrite grain size and the prior austenite grain size, respectively. In the case of QT,
the d_{EFF} value of 25 μm fitted with the martensite packet size in the tempered

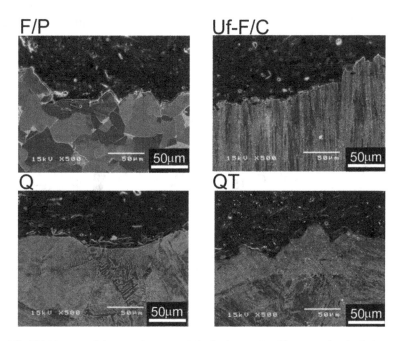

Fig. 3.7 SEM images of cleavage fracture unit for ferrite/pearlite (F/P), ultra-fine ferrite/cementite (UGF/C), as-quenched (Q), and quench-and-tempered (QT) steels (after [1])

structure. The reason for the difference in the d_{EFF} values of Q and QT might be the relaxation of strain and the rotation of crystallographic grain direction because of annealing. The d_{EFF} value of 8 μm for UGF/C corresponded to the larger grain size in the bimodal distribution.

Figure 3.8 shows the relationship between d_{EFF} and DBTT, including the literature data for QT steels [12], F/P steels (0.15C-0.26Si-1.44Mn) [13], and another F/P steel (SM490: 0.15C-0.48Si-1.48Mn) [19]. The F/P steel of Ref. [19] showed an average grain size of 7 μm and a DBTT of 200 K. Here, the data of the two F/P samples with different grain sizes of 12 and 16 μm were also added. These two samples were prepared by heating the UGF/C sample at different temperatures in the austenite phase region and then by cooling them in the F/P structure. In this figure, two groups seemed to exist, one belonging to the F/P microstructure and the other to the Q, QT, and UGF/C microstructures.

The important implication of Fig. 3.8 is that the smallest d_{EFF} of UGF/C results in the lowest DBTT in the latter group. Furthermore, the UGF/C showed a lower DBTT at a given d_{EFF} than the F/P group. When d in Eq. (3.2) was replaced by d_{EFF}, DBTT can be, accordingly, expressed by the following equation:

Fig. 3.8 Relationship
between effective grain size
and DBTT for the steels in
this study, M and M + B
steels [9] and an F/P steel [10,
16] (after [1])

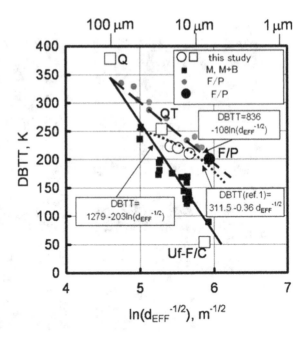

$$DBTT = A - B \cdot \ln(d_{EFF}^{-1/2}) \qquad (3.3)$$

Here, A and B are 1,279 and 203 for the Q, QT and UGF/C group, and 836, 108 for the F/P group. The unit of DBTT is K for DBTT and that of d_{EFF} is m.

In Fig. 3.8, the estimation on the change of DBTT by d_{EFF} is also superimposed using Eq. (3.1), which was proposed by Pickering [3].

$$DBTT = 254 + 44(Si) + 700(N_f)^{-1/2} + 2.2(Pearlite) - 0.36\,d^{-1/2} \qquad (3.4)$$

with Si = 0.3 %, N_f = 0 %, Pearlite = 20 %. Finally, A becomes 311.5 and B 0.36 in the form of Eq. (3.1). Here (Si) and (Nf), are the mass percentages of Si and N in the matrix, respectively. (Pearlite) is the volume percent of pearlite. The unit of d is m. The estimated curve was close to the F/P group data in the present study. However, the extrapolated curve showed that the DBTT of F/P was lower than that encountered in the present study, suggesting that Pickering's estimation is applicable only in the range around 10 μm. For grain sizes between 100 and 7 μm, the equation obtained in this study along with the data from Refs. [13] to [19], e.g. Eq. (3.3) with A = 836 and B = 108, agreed better with the presented data.

3.1.3 Estimation of σ_F

We know that σ_F is a function of the surface energy of fracture γ and the crack size r, according to the Griffith equation [20] when considering an infinite cracked plate with a central transverse crack of length r:

$$\sigma_F = (4\gamma E/\pi)^{1/2} \cdot r^{-1/2} \tag{3.5}$$

Here we replace r by d_{EFF} to obtain Eq. (3.6).

$$\sigma_F = (4\gamma E/\pi)^{1/2} \cdot d_{EFF}^{-1/2} \tag{3.6}$$

However, when we assume an infinite plate with a penny shaped crack of diameter r in the bulk center, the fracture stress is expressed as follows [20]:

$$\sigma_F = (\pi\gamma E/(v-1)^2)^{1/2} \cdot r^{-1/2} \tag{3.7}$$

Here, we replace r by d_{EFF}:

$$\sigma_F = (\pi\gamma E/(v-1)^2)^{1/2} \cdot d_{EFF}^{-1/2} \tag{3.8}$$

When we assume that σ_F is independent of temperature and becomes equal to the flow stress at DBTT, it is possible to estimate σ_F by taking YS as the flow stress at DBTT, as described below.

According to Tsuchida et al. [21], the thermal component of flow stress is independent of the ferrite grain size, while the athermal component increases with decrease in the grain size. Therefore, in Fig. 3.9, the change in the flow stress by temperature obtained by Tsuchida et al. was adopted and drawn parallel to each sample as each curve consisted of YS determined at three different temperatures:

Fig. 3.9 Relationship between flow stress and σ_F as a function of temperature based on experimental data for different microstructures or in other words the estimated σ_F (after [1])

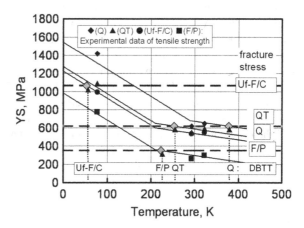

77, 293, and 323 K. However, in sample Q, the curve was offset to higher temperature regions to fit the experimental data of YS. This was done by assuming that the change in YS for sample Q has the same shape as the F/P samples, at a different temperature, at which the curve changes its inclination. The marks ■, ●, ◆, and ▲ stand for the YS of F/P, UGF/C, Q, and QT, respectively. The large meshed diamond (◈) is the YS at DBTT for each sample, i.e., the estimated σ_F. The horizontal line shows the σ_F versus temperature curve for each sample. The difference between σ_F and YS for each microstructure was 440, −18, and −90 MPa at room temperature for UGF/C, Q, and QT, respectively, while F/P was taken as the reference. This clearly showed that UGF/C showed the largest difference between σ_F and YS among the four microstructures. Hence, the earlier speculation that the grain refinement significantly increases the σ_F value is proven.

3.1.4 Estimation of γ

It is also possible to estimate γ for each sample with the estimated σ_F and the experimentally determined d_{EFF} by Eqs. (3.6) and (3.8). In this estimation, E is taken as 2.06×10^2 GPa and n is taken as 0.293 for steel using Eqs. (3.6) and (3.8).

When σ_F determined by Eq. (3.6) was plotted against d_{EFF}, we could find three similar groups as in Fig. 3.8. The slope of each group showed γ. The first group consisted of only the F/P samples with the lowest estimated surface energy of 7.7 J/m², the second group consisted of UGF/C and QT with an estimated surface energy of 34.6 J/m², and the third group consisted of Q with a surface energy of 150.9 J/m². The value of σ_F determined by Eq. (3.8) could also be used to estimate the surface energy of each sample, and we could find three groups. In this case, the first group consisted of only the F/P samples with the lowest estimated surface energy of 1.5 J/m², the second group of the UGF/C and QT samples with an estimated surface energy of 7.0 J/m², and the third group consisted of the Q samples with a surface energy of 29.7 J/m².

The surface energy of ferrite iron has been shown to be 4.52 J/m² by Kelly et al. [22]. Yokobori et al. [23] reported the experimental value of the effective surface energy as 100 J/m². In Fig. 3.10, σ_F determined by Eq. (3.6) is plotted against d_{EFF}, together with lines constructed using the value of surface energy shown in Refs. [22, 23]. The surface energy together with d_{EFF}, and the unit of the structure determining the d_{EFF}, YS, σ_F, and DBTT are listed in Table 3.3. It is interesting to note that UGF/C shows a higher surface energy than F/P. This means that UGF/C can have a lower DBTT than F/P because of its higher surface energy of fracture and the resultant higher fracture stress when UGF/C has the same d_{EFF} as that of F/P. Accordingly, the excellent toughness or the very low DBTT of UGF/C can be attributed to its small d_{EFF} and high γ.

Fig. 3.10 σ_F as a function of the inverse square root of effective grain size for different microstructures (after [1])

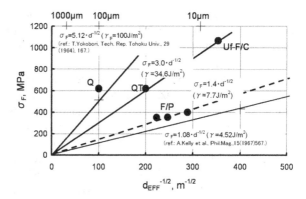

Table 3.3 Properties of steel with different microstructures, i.e., ferrite/pearlite (F/P), ultra-fine ferrite/cementite (UGF/C), as-quenched (Q), and quench-and-tempered (QT) steels: d_{EFF} (effective grain size), unit determining the d_{EFF}, YS (yield strength), σ_F (fracture strength), DBTT (ductile-to-brittle transition temperature), and (surface energy) (after [1])

Sample	d_{EFF} (um)	Unit determining d_{EFF}	YS (MPa) experimental	σ_F (MPa) estimated	DBTT (K) experimental	T (J/m^2) estimated
F/P	20	Ferrite grain	261	351	223	7.7
UGF/C	8	Bigger part in grain clusters	536	1,066	53	34.6
QT	25	Martensite packet	548	620	253	
Q	100	Prior-austenite grain	619	619	378	150.9

3.1.5 Bimodal Distribution of Grain Size in UGF/C

The present UGF/C sample had three types of grain sizes: 1.6 and 8 μm for the peaks of distribution, and 2.7 μm for the area average determined by the EBSD analysis. One method to determine the grain size of a microstructure is by the linear intercept method based on the OM or SEM images. Another method is the grain area averaging method based on the EBSD images using different angle values of grain boundaries. Among the high angle grain boundaries, those with boundary angles equal to or above 5° are said to be effective in clarifying the grain sizes [24, 25]. In the UGF/C sample, the larger grain cluster, i.e., the 8-μm grain cluster, corresponded to the d_{EFF} value.

Tsuchida et al. [26] reported that YS is estimated to be 551 MPa for 1.6 μm, 464 MPa for 2.7 μm, and 342 MPa for 8-μm grains when UGF/C is annealed after caliber-rolling. Yin et al. [17] considered the effect of average dislocation density and the dislocation interaction parameter on YS and concluded that the dislocation contribution is estimated to be 115 and 52 MPa in the as-rolled and as-annealed conditions, respectively. Based on this, the dislocation density contribution without annealing is estimated to be 63 MPa. When this value of 63 MPa is subtracted from the YS data for the presently considered UGF/C sample, the corrected YS value for the annealed sample became 473 MPa. This was in good agreement with the above-mentioned value of 464 MPa, because the grain size was supposed to be 2.7 μm.

Therefore, the grain size that determined the YS is the average grain size for the UGF/C sample, although the d_{EFF} value that determines the DBTT corresponded to the larger size in the two-grain clusters. This result might be highly disputable; however, more improvement in the YS-DBTT balance can be expected from the present results. Further detailed studies are required for this purpose.

To investigate the excellent toughness of UGF/C steels, the concept of d_{EFF} was applied to DBTT for four different microstructures in a single composition of low-carbon steel. The microstructures were UGF/C, F/P, Q, and QT. The following conclusions were obtained:

(1) For UGF/C, F/P, Q, and QT, the d_{EFF} values were determined to be 8, 20, 100, and 25 μm, respectively. In F/P and Q, the d_{EFF} value was in accordance with the ferrite grain size and the prior austenite size, respectively. The d_{EFF} value for QT corresponded to the martensite packet size. In UGF/C, the ferrite grain size showed a bimodal distribution and the larger grain size matched with d_{EFF}, the smallest among the four microstructures.

(2) The relationship between d_{EFF} and DBTT showed that the UGF/C micro-structure belonged to the same group composed of Q and QT microstructures, while F/P belonged to a different group.

(3) The estimated fracture stress showed that the UGF/C sample has the highest fracture stress among the four microstructures.

(4) The difference in DBTT and σ_F among the four microstructures was estimated to be the difference in γ. The group of UGF/C and QT showed a much higher surface energy value than the F/P group.

(5) The excellent toughness of the UGF/C steel could be attributed to its char-acteristic small d_{EFF} and high γ compared to the other structures.

3.2 Low Absorbed Energy Ductile Dimple Fracture in Lower Shelf Region in UGF/C Steel

The curve of absorbed energy versus test temperature for the UGF/C steel shows a transition from upper shelf energy to lower shelf energy, i.e., a transition from an energy-absorbent ductile mode to an energy-absorbent brittle mode. Some dense and small dimples were observed in the lower shelf region. A significant and consistent difference existed between DBTT and impact energy transition temper-ature, i.e., the steel had low absorbed energy but underwent a ductile dimple fracture at a certain low temperature (below around 100 K).

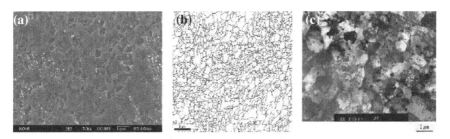

Fig. 3.11 Characteristics of the UGF/C microstructure: **a** SEM image, **b** EBSD map, and **c** TEM image (after [2])

Ultra-fine-grained C–Mn steels with average ferrite grain sizes smaller than a few microns and even smaller than 1 μm have been recently successfully developed and are generally characterized by an ultra-fine-grained ferrite matrix plus finely globular cementite particles (hereafter, UGF/C steels) [27–29]. The performance of steel in structural applications is generally restricted by its impact toughness. In the engineering approach, the impact toughness is often quantified in relation to the Charpy impact test in which the steels show a transition from high absorbed energy with a ductile dimple fracture at high temperatures to low absorbed energy with a brittle cleavage fracture at low temperatures [30, 31]. In this work, the Charpy impact test is conducted on the UGF/C steel. Special attention was given to the absorbed energies and the appearances of the broken impact pieces in the lower shelf region.

The UGF/C microstructure was obtained in steel with a chemical composition (wt%) of 0.154C, 0.301Si, 1.504Mn, 0.011Al, 0.002S, and 0.001P by caliber warm-rolling [32]. The ingots were melted in vacuum in a laboratory scale, homogenized for 60 min at 1,473 K, hot-forged into 115-mm diameter rods, and subjected to multi-pass caliber warm-rolling to obtain the final 17-mm diametral rods by an accumulated area reduction of 95 %.

The microstructure was analyzed in the transverse direction by SEM, EBSD, and TEM. Standard 10-mm square, 55-mm long, and 2-mm V-notched geometrical Charpy impact test pieces were prepared. The length was the same as that of the rod. The Charpy impact tests were conducted at temperatures ranging from 4.2 to 373 K, and the fracture surfaces of the broken pieces were examined with a digital camera and an SEM. The quantitative analysis of the dimple size was performed on at least five SEM images for each condition.

An SEM image of the UGF/C microstructure is shown in Fig. 3.11a. The ultra-fine-grained ferrite matrices were heterogeneously populated by numerous globular cementite particles. These particles are less than 100 nm in diameter and were distributed with a local high density. An EBSD map for the UGF/C microstructure is shown in Fig. 3.11b, in which the ferrite grains were manifested by the boundaries with misorientations above 1°. A TEM image of the UGF/C microstructure is shown in Fig. 3.11c. Figure 3.11a–c show that the microstructure was characterized by ultrafine ferrite grains and cementite particles and was a typical

Fig. 3.12 a Curve of
absorbed energy versus test
temperature, showing a
transition from an *upper* shelf
region to a *lower* shelf region.
b Fracture surfaces of the
broken impact pieces at
various test temperatures
(after [2])

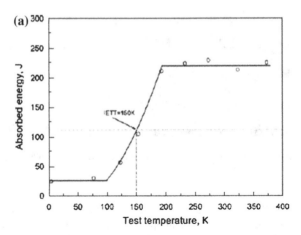

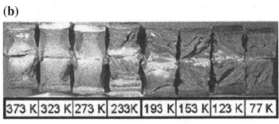

UGF/C microstructure. The average ferrite grain size was about 0.7 μm. The
observation of the absorbed energy with the test temperature is illustrated in
Fig. 3.12a, with a transition from the upper shelf temperature region (≥193 K) to the
lower shelf temperature region (≤77 K).

The absorbed energy revealed a gradual transition from the upper shelf energy
(220 J) to the lower shelf energy (25 J) with decrease in the test temperature. The
impact energy transition temperature (IETT), which is the temperature that corresponds to half the value of the upper shelf energy, is determined to be 150 K. A
series of fracture photographs of the broken impact pieces is shown in Fig. 3.12b.
Each fracture surface in the upper shelf region of 373, 323, and 273 K is rough and
obviously ductile with macroscopic plastic deformation. However, the separation
began to occur at the upper shelf temperature regions of 233 and 193 K. The
absorbed energies at these temperatures did not change much regardless of the
fracture surface appearances. The separation also occurred at test temperatures
below 193 K, i.e., in the transition temperature and lower shelf temperature regions.

The appearances of the broken impact pieces at the representative temperatures
are presented in Fig. 3.13. The area at the center of the piece and 2 mm below the
V-notch was observed. The adjoining and large dimples and the dense and small
dimples were observed on the fracture surface tested at 323 K in the upper shelf
region. The former consisted of an average diameter of approximately 12 μm and an
area fraction of 30 percent (Fig. 3.13a). The adjoining and large dimples decreased

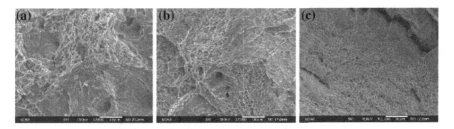

Fig. 3.13 Fracture surfaces of broken impact pieces: **a** 323 K, **b** 153 K, and **c** 77 K (after [2])

in both the area fraction and size at 153 K in the transition region, but the dense and small dimples increased in the area fraction with little change in size (Fig. 3.13b). In addition, the adjoining and large dimples disappeared almost completely and showed the appearance of some cleavages (approximately 20 % in area fraction), while the dense and small dimples still remained with little change in size at 77 K in the lower shelf region (Fig. 3.13c). The size and distribution of dimples are summarized in more detail in Fig. 3.14a. This characteristic change in the size and distribution of dimples can be expected to affect the transition from the upper shelf energy to the lower shelf energy.

The change in the crystallinity or the percentage of cleavage fracture area on the fracture surface is shown in Fig. 3.14b and is plotted as a function of the test temperature. The DBTT is defined here as the temperature with 50 % crystallinity. No DBTT was obtained even at a test temperature as low as that of liquid helium. Interestingly, there is a significant and consistent difference between DBTT (less than 4.2 K) and IETT (150 K), i.e., the low absorbed energy while undergoing a ductile dimple fracture at a certain low temperature.

The fracture surfaces in the lower shelf region revealed numerous dense and small dimples, rather than the conventionally expected complete cleavage or quasi-cleavage fracture. Steels are generally accompanied by a ductile dimple fracture to a brittle cleavage fracture change at a certain low temperature [33]. This can be quantitatively interpreted as caused by the competition between the flow stress and σ_F [34, 35]. The flow stress increases with decrease in temperature, whereas σ_F is independent of temperature [35].

The fracture surfaces will undoubtedly present dimples at the temperature region where σ_F is higher than the flow stress. This is because plastic deformation prevails over brittle fracture when yielding occurs first, while revealing cleavages at the temperature regions when σ_F is lower than the flow stress. This is because of the fact that the cleavage starts before yielding and without any plastic deformation. The dense and small ductile dimples still remain in the lower shelf region at a test temperature as low as that of liquid helium (4.2 K), implying the occurrence of a ductile fracture to a certain extent.

Numerous nano particles were observed in the dense and small dimples in high magnification (Fig. 3.15a). These extremely fine particles are generally not revealed easily in chemical compositions. However, cementite particles were the most

Fig. 3.14 a Size and area
fraction of different types of
dimples as a function of test
temperature. **b** Crystallinity or
area percent of cleavage
fracture on fracture surface
(after [2])

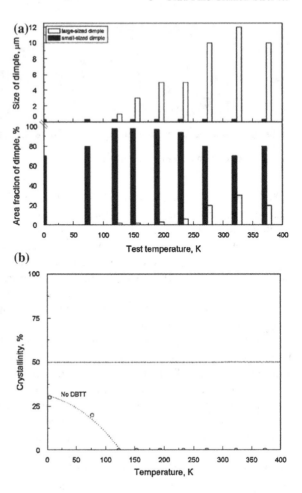

prevalent whilst considering the size and the distribution of cementite in the
microstructure.

A side view of the fracture surface of the impact piece tested at 77 K is shown in
Fig. 3.15b. Some microscopic voids were observed at the site with a local high
density of cementite particles. One possible mechanism for the ductile dimple
fracture in this circumstance is that slip bands impinge on the cementite particles
causing local strain concentrations where the voids are nucleated. A small scale
ductile rupture along the cementite particles is then believed to cause the failure.

Ultra-grain refinement of ferrite is another possible mechanism contributing to
this characteristic ductile fracture, because it essentially decreases the DBTT [35].
Detailed discussion on the mechanism that governs this characteristic ductile
fracture should be studied in the future. It is believed that low absorbed energy with
a ductile dimple fracture in the lower shelf region is a characteristic feature of the
UGF/C steel.

Fig. 3.15 a Dense and small-sized dimples embedded with nanosized particles. **b** Side view of the fracture surface of the broken impact piece, showing voids at the site with local high density of cementite particles (the underside of the specimen) (after [2])

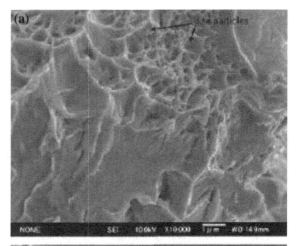

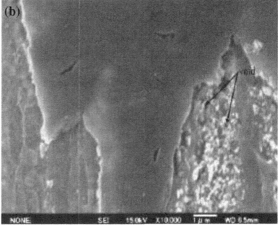

Based on the discussion in this chapter, we can arrive at the following conclusions:

(1) In the ultra-fine-grained steel, a transition from an energy-absorbent ductile mode to an energy-absorbent brittle mode existed. Some dense and small dimples were observed in the lower shelf region.

(2) A significant and consistent difference existed between the DBTT and IETT in the ultra-fine-grained steel, i.e., low absorbed energy while undergoing a ductile dimple fracture at a certain low temperature.

(3) Dense and small ductile dimples still remained in the lower shelf region at a test temperature as low as 4.2 K in the ultra-fine-grained steel, implying the occurrence of a ductile fracture.

References

1. T. Hanamura, F. Yin, K. Nagai, Ductile-brittle transition temperature of ultrafine ferrite/cementite microstructure in a low carbon steel controlled by effective grain size. ISIJ Int. **44**, 610–617 (2004)
2. M. Zhao, T. Hanamura, H. Qui, H. Dong, K. Yang, K. Nagai, Low absorbed energy ductile dimple fracture in lower shelf region in an ultrafine grained ferrite/cementite steel. Metall. Mater. Trans. A **37A**(9), 2897–2900 (2006)
3. F.B. Pickering, T. Gladman, Metallurgical developments in carbon steels. ISI Spec. Rep. **81**, 10 (1963)
4. A.H. Cottrell, Theory of brittle fracture in steel and similar metals. Trans. Metall. Soc. **212**, 192 (1958)
5. N.J. Petch, The ductile to brittle transition in the fracture of alpha-iron. Philos. Mag. **3**, 1089 (1958)
6. A.N. Stroh, A theory of the fracture of metals. Adv. Phys. **6**(24), 418 (1957)
7. T. Hanamura, T. Hayashi, H. Nakajima, S. Torizuka, K. Nagai, in *Second International Conference on Processing Materials for Properties*, ed. by B. Mishra, C. Yamauchi (TMS, Warrendale, 2000), p. 206
8. W.C. Leslie, Iron and its dilute substitutional solid solutions. Metall. Trans. **3**, 5–26 (1972)
9. J.M. Hodge, R.D. Manning, H.M. Reichhold, The effect of ferrite grain size on notch toughness. Trans. AIME **185**, 185–233 (1949)
10. S. Matsuda, T. Inoue, M. Ogasawara, The fracture of tempered martensite. Trans. Jpn. Inst. Met. **9**, 343 (1968)
11. S. Matsuda, T. Inoue, M. Ogasawara, The fracture of a low carbon tempered martensite. Trans. Jpn. Inst. Met **11**, 36 (1970)
12. F. Terasaki, H. Ohtani, The microstructure and toughness of high tensile strength steels. Tetsu-to-Hagané **58**, 436 (1972)
13. F. Terasaki, H. Ohtani, Trans. Iron Steel Inst. Jpn. **12**, 45 (1972)
14. C.D. Beachem, Orientation of cleavage facets in tempered martensite (quasi- cleavage) by single surface trace analysis. Metall. Trans. **4**, 1999–2000 (1973)
15. S. Matsuda, T. Inoue, H. Mimura, Y. Okamura, Trans. Iron Steel Inst. Jpn. **12**, 325 (1972)
16. C.D. Beachem, *Fracture*, vol. I (Academic Press, New York, 1968), p. 305
17. F. Yin, T. Hanamura, T. Inoue, K. Nagai, Characteristic microstructure features influencing the mechanical behavior of warm-rolled ultrafine low-carbon steels, in *Seventh Workshop on the Ultra-Steel: Requirements from New Design of Constructions*, NIMS, Tsukuba, p. 288 (2003)
18. F. Terasaki, H. Ohtani, Tetsu-to-Hagané **58**, 1067 (1972)
19. K. Nagai, O. Umezawa, Report of study committee on effect of impurities due to scrap on steel products, Materials Science of Tramp Elements, ISIJ, Tokyo, p. 70 (1997)
20. A.A. Griffith, Philos. Trans. R. Soc. (London) A, **A221**, 163 (1920)
21. N. Tsuchida, Y. Tomoda, K. Nagai, ISIJ Int. **42**, 1594 (2002)
22. A. Kelly et al., Philos. Mag. **15**, 567 (1967)
23. T. Yokobori, Tech. Rep. Tohoku Univ. **29**, 167 (1964)
24. H. Yagi, N. Tsuji, Y. Saito, Tetsu-to-Hagané **86**, 349 (2000)
25. A. Ohmori, S. Torizuka, K. Nagai, K. Yamada, Y. Kogo, Tetsu-to-Hagané **88**, 857 (2002)
26. N. Tsuchida, Y. Tomoda, K. Nagai, Tetsu-to-Hagané **89**, 1170 (2003)
27. K. Nagai, Ultrafine-grained ferrite steel with dispersed cementite particles. J. Mater. Proc. Technol. **117**(3), 329–332 (2001)
28. L. Storojeva, D. Ponge, R. Kaspar, D. Raabe, Development of microstructure and texture of medium carbon steel during heavy warm deformation. Acta Mater. **52**(8), 2209–2220 (2004)
29. R. Song, D. Ponge, D. Raabe, R. Kaspar, Microstructure and crystallographic texture of an ultrafine grained C–Mn steel and their evolution during warm deformation and annealing. Acta Mater. **53**(3), 845–858 (2005)

30. L. Toth, P. Rossmanith, Historical background and evaluation of the Charpy test, in *Charpy Centenary Conference*, Poitiers, France, 2–5 October 2001, p. 1. European Structural Integrity Society, Paris (2001)

31. T. Gladman, *The Physical Metallurgy of Microalloyed Steels* (The Institute of Materials, London, 2002), p. 57

32. M. Zhao, T. Hanamura, H. Qui, K. Nagai, K. Yang, Dependence of strength and strength-elongation balance on the volume fraction of cementite particles in ultrafine grained ferrite/cementite steels. Scripta Mater. **54**(7), 1385–1389 (2006)

33. G. Krauss, *Steels Heat Treatment and Processing Principles* (ASM International, Materials Park, 1990), pp. 133–139

34. F.B. Pickering, T. Gladman, Iron and Steel Institute Special Report No. 81, Iron and Steel Institute, Tokyo, p. 10 (1963)

35. T. Hanamura, F. Yin, K. Nagai, Iron Steel Inst. Jpn. Int. **44**, 610–617 (2004)

Chapter 4
Ultra-Fine-Grained Steel: Fracture Toughness (Crack-Tip-Opening Displacement)

Toughness represents the resistance to fracture. Let us suppose that a sharp crack exists in a material, as illustrated in Fig. 4.1. When an applied stress is loaded, the initial sharp crack will open and plastic deformation begins to occur ahead of the crack tip. The plastic deformation blunts the crack tip, and the degree of crack blunting enlarges as the applied stress increases. The opening displacement at the crack tip (δ), as illustrated in Fig. 4.1, is denoted as crack-tip-opening displacement (CTOD). CTOD increases as the applied stress increases, and when a certain value is reached, the crack will begin to propagate. CTOD at this point is referred to as the critical CTOD. If crack propagation takes place, it is generally regarded as a failure of the material, and thus the toughness at the beginning of crack propagation is crucial. The critical CTOD is a measure of resistance against the beginning of crack propagation and is used as a parameter of fracture mechanics to evaluate the material's fracture toughness. Critical CTOD increases in proportion to the material toughness.

A crack propagates in different fracture modes in line with the material toughness. There are two typical fracture modes in ferritic steels: ductile fracture and brittle fracture, with the former corresponding to high toughness and the latter to low toughness. The occurrence of crack propagation obeys certain criteria in association with fracture modes.

Figure 4.2 illustrates the conditions for crack propagation in brittle and ductile materials [1]. Let us suppose that a stationary crack is subjected to a load of mode I (i.e., the opening mode). The opening tensile stress (σ_{yy}) distribution and plastic strain distribution directly ahead of the crack tip along crack direction can be schematically represented as shown in Fig. 4.2. Because of the presence of cracks, in addition to the stress and strain concentration, stress singularity and strain singularity exist near the crack tip.

It is known that ductile fracture and brittle fracture are strain-controlled and stress-controlled, respectively [2]. The critical condition for the initiation of brittle fracture (or ductile fracture) is that over a characteristic distance l_0, as shown in Fig. 4.2, the local tensile stress perpendicular to the crack plane σ_{yy} (or local

© National Institute for Materials Science, Japan. Published by Springer Japan 2014   49
T. Hanamura and H. Qiu, *Analysis of Fracture Toughness Mechanism in Ultra-fine-grained Steels*, NIMS Monographs, DOI 10.1007/978-4-431-54499-9_4

Fig. 4.1 Crack tip opening displacement (CTOD)

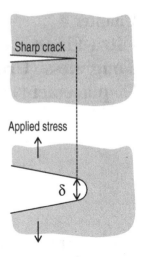

Fig. 4.2 Schematic diagrams of microscopic fracture criteria for brittle fracture and ductile fracture (after [1])

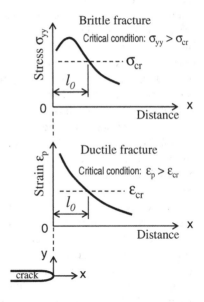

equivalent plastic strain ε_p) exceeds the critical fracture stress σ_{cr} (or the critical fracture strain ε_{cr}) [2]. The mode in which a material fractures is expected to depend on the type of critical condition first attained. For example, if the condition for ductile fracture is first satisfied, ductile fracture occurs, otherwise, brittle fracture takes place.

A schematic diagram interpreting the correlation of the stress–strain curve with the fracture conditions (σ_{cr}, ε_{cr}) for all fracture modes is given in Fig. 4.3. The stress–strain curves represent the relationship between σ_{yy} and ε_p at the point ahead of the crack tip along the crack direction within the characteristic distance l_0 shown in Fig. 4.2, where the stress or strain is the minimum.

Fig. 4.3 Schematic interpretation of the effect of grain refinement on fracture mode. *Solid* and *dashed lines* corresponding to the grain size of D_i and D_{i+1}, respectively ($D_i > D_{i+1}$) (after [1])

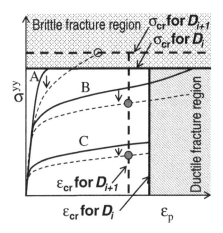

Three typical cases (solid lines, corresponding to the ferrite grain size D_i) are shown in Fig. 4.3. In Case A and Case C, as the stress and strain increase, only one critical value (σ_{cr} in Case A and ε_{cr} in Case C) is exceeded; accordingly, the fracture modes for Case A and Case C are the brittle and ductile modes, respectively. In Case B, the ductile fracture condition is first satisfied and is subsequently followed by the exceeding of the critical fracture stress. Thus, the material first fractures in a ductile mode and then in a brittle mode.

For a given crack, stress or strain concentration is related to the material's work-hardening ability. For plane strain state, altering the work-hardening exponent (n) will change the local stress field ahead of a crack tip. If we take remote flow stress σ_0 as $\sigma_0/E = 0.0025$ (where E is the Young's modulus), the local peak stress (as shown in Fig. 4.2) can reach $\sim 3\sigma_0$, $\sim 3.6\sigma_0$, and $\sim 5\sigma_0$ for $n = 0, 0.1$, and 0.2, respectively [2]. This result shows that decreasing the work-hardening exponent (i.e., the work-hardening ability) reduces the level of local stress ahead of a crack. As for the corresponding near-tip equivalent plastic strain distribution, it is essentially independent of the work-hardening ability [3].

Besides the stress and strain fields ahead of the crack tip, the σ_{cr} and ε_{cr} values are also strongly related to the characteristics of the material. Because grain refinement greatly affects the characteristics of the material, the effect of grain refinement should be investigated. First, the effect of grain refinement on the work-hardening ability is presented.

Let us assume that the true stress (σ)-true strain (ε) curves obey the Hollomon equation (Eq. 4.1).

$$\sigma = K\varepsilon^n \tag{4.1}$$

where K is a constant and n is referred to as the work-hardening exponent. In the engineering field, n is often used to express the degree of work-hardening: the larger the n value is, the greater is the work-hardening ability. By fitting the

Fig. 4.4 Effects of ferrite grain size and carbon content on the work-hardening exponent (after [4])

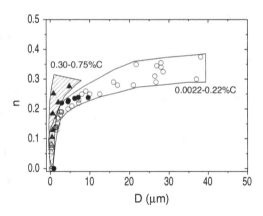

experimental σ-ε curves obtained from tensile tests with Eq. (4.1), the n value can be obtained.

Figure 4.4 shows the dependence of work-hardening exponent (n) on the ferrite grain size (D) for ferritic steels with carbon content 0.0022–0.75 % [4]. The n value decreases as D reduces. This demonstrates that grain refinement lowers the work-hardening ability. Morrison proposed an empirical equation for the relation between n and ferrite grain size D as follows [5]:

$$n = \frac{5}{10 + D^{-1/2}} \qquad (4.2)$$

where D is in mm. Equation (4.2) indicates that a decrease in D produces a small n value and n is a function of only D. However, apparently, the data in Fig. 4.4 can be roughly classified into two groups according to the carbon content, with the higher carbon content steel having an elevated n value for a given D. This result demonstrates that Eq. (4.2) cannot exactly express Fig. 4.4, and the effect of carbon (i.e., cementite, herein) should be taken into account.

Antoine et al. [6] recognized that all the factors controlling the yield strength will also affect n. Further, they provided an equation for n derived from the relationship between the yield strength and n within the ferrite of sizes ranging from 21.7 to 38.1 μm for IF steel with ferrite/precipitates ($Ti_4C_2S_2$, TiFeP, TiC). By using their approach, n was plotted against the yield strength in Fig. 4.5. To enlarge the range of experimental data and increase the statistical accuracy, a vast amount of data is summarized. A linear behavior for yield strength was observed for n values when the carbon content was below 0.22 %; however, as the carbon content further increased, n values significantly deviated from the linear behavior. The regression equation for C ≤ 0.22 % is as follows:

Fig. 4.5 Correlation between yield strength and work-hardening exponent (after [4])

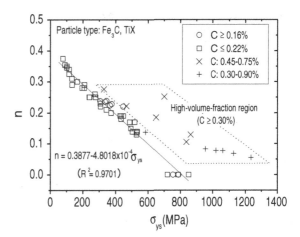

$$n = 0.3877 - 4.8018 \times 10^{-4}\sigma_{ys} \qquad (\sigma_{ys} \text{ in MPa}) \qquad (4.3)$$

The square of the correlation coefficient, R^2, for Eq. (4.3) is 0.9701.

Ohmori et al. [7] summarized the relationship between the lower yield strength and the ferrite grain size with a significant amount of data on ferrite/cementite steels with carbon content varying from 0.10 to 0.3C, and provided Eq. (4.4).

$$\sigma_{ys} = 120 + 500D^{-1/2} \qquad (4.4)$$

where the ferrite grain size D is in μm.

Substitution of Eq. (4.4) into Eq. (4.3) yields the following equation:

$$n = 0.3301 - 0.2401D^{-1/2} \qquad (D \text{ in } \mu m) \qquad (4.5)$$

The experimental data for ferrite/cementite steels (carbon content: 0.05–0.22 %; ferrite grain size: 0.25–37 μm) are compared with Eqs. (4.2) and (4.5) in Fig. 4.6. The square of the correlation coefficient, R^2, for particle sizes ranging from 1.30 to 37 μm is 0.9359 for Eq. (4.5) and 0.9153 for Eq. (4.2), which shows that the accuracy of Eq. (4.5) is better than that of Eq. (4.2).

Equation (4.5) demonstrates that n continuously declines with a decrease in the ferrite grain size; when $D = 0.53$ μm, n equals zero, i.e., work-hardening disappears for grain sizes below 0.53 μm. This phenomenon is principally attributed to the decreased number of dislocation sources and impaired/reduced storage of dislocations. The linear relation between n and $D^{-1/2}$ was confirmed within the range 1.30–37 μm. Beyond this range, the experimental data is insufficient to draw any conclusion.

The critical grain size for an n value given by Eq. (4.5) for $n = 0$ is 0.53 μm. However, the experimental data shows that n already becomes zero when

Fig. 4.6 Prediction of the work-hardening exponent using Eqs. (4.2) and (4.5) (after [4])

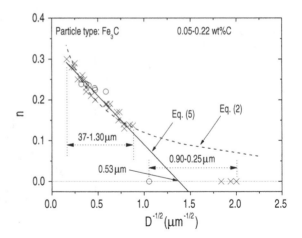

$D = 0.9$ μm, indicating that the error of Eq. (4.5) is large in the vicinity of $n = 0$. Therefore, Eq. (4.5) is valid for ferrite grain sizes of 1.30–37 μm, and the requirement for metallurgical conditions such as carbon content and cementite particle size are the same as those for Eq. (4.3).

The dependence of work-hardening on grain size and cementite volume fraction around the boundary of $n = 0$ is summarized in Fig. 4.7. The experimental data are partitioned by a dashed line that delineates work-hardening and the absence of work-hardening above/below the line, respectively. The partition line shows the relationship of the ferrite grain size D with cementite volume fraction f_v for the critical condition $n = 0$. Because of insufficient data, the critical partition line is inexact, particularly at low f_v, in Fig. 4.7. When f_v is larger than 4.58 %, even if D is very small; e.g., 0.25 μm, n is still not equal to zero because of the contribution of cementite particles to work-hardening. If f_v is sufficiently small, work-hardening will disappear at a relatively larger grain size. For conventional ferrite/cementite

Fig. 4.7 Influence of ferrite grain size and cementite volume fraction on the critical condition of $n = 0$ (after [4])

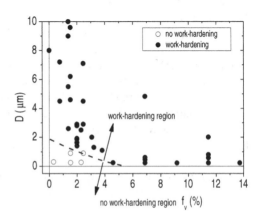

steels (C 0.22 %, i.e., f_v 3.4 %), there should be no work-hardening ability when D is less than 1 μm.

From the above discussion, it can be understood that grain refinement lowers and even completely nullifies the work-hardening ability. Therefore, when the ferrite grain size reduces from D_i to D_{i+1} ($D_i > D_{i+1}$), the n value will decrease. As discussed above, decreasing n reduces the local stress concentration ahead of a crack; however, this has little effect on the local strain distribution. Therefore, the decrement in n induced by grain refining from D_i to D_{i+1} merely lowers the local stress, and as a result, the three solid lines will drop, as shown in Fig. 4.3 (from solid lines to dashed lines).

Cottrell [8] analyzed the critical fracture stress for transgranular cleavage with dislocation theory and found that the critical fracture stress σ_{cr} is proportional to $D^{-1/2}$. For steels and irons, the relationship between σ_{cr} and D in the range 10–10,000 μm is given in the form of Eq. (4.6) [9]:

$$\sigma_{cr} = 351.5 + 3.30 \times 10^3 D^{-1/2} \qquad (D \text{ in } \mu m; \sigma_{cr} \text{ in MPa}) \qquad (4.6)$$

Therefore, the grain refinement elevates σ_{cr}, as shown in Fig. 4.3.

To investigate the dependence of critical fracture strain on the ferrite grain size, tensile tests were performed at room temperature and at a crosshead rate of 0.4 mm/min with specimens of 4 mm diameter and 10 mm gage length. The steels and samples used were (1) 0.10C-0.3Si-1.45Mn (in wt%) ferritic steels, grain size: 0.9, 4.6, 6.2, and 9.6 μm; (2) 0.16C-0.3Si-1.45Mn ferritic steels, grain size: 0.9, 2.9, 4.5, and 7.1 μm; (3) 0.45C-0.3Si-1.45Mn ferritic steels, grain size: 0.46, 0.60, and 4.83 μm; (4) 0.75C-0.3Si-1.45Mn ferritic steels, grain size: 0.58, 0.80, and 2.01 μm. All the samples were composed only of ferrite and second phase particles (cementite particles). The fracture strain of the tensile specimens, ε_f, was given by the following equation:

$$\varepsilon_f = \ln(A_0/A_f) \qquad (4.7)$$

where A_0 and A_f are the areas of the initial and the fractured cross sections, respectively. The results are given in Fig. 4.8. The results show that decreasing the ferrite grain size reduced the fracture strain. The carbon content, i.e., the cementite volume fraction, also affects the fracture strain; higher carbon content is harmful to fracture strain.

Round smooth tensile specimens were applied in Fig. 4.8; thus, before the onset of necking, the specimens were in a uniaxial stress state, while after necking the stress state turned into a triaxial state. For a sample with a crack, because of the crack tip sharpness, the stress triaxiality in the region ahead of the crack tip was higher than that in the round smooth tensile specimen. Therefore, the fracture strain obtained from tensile tests was usually larger than the critical fracture strain in the local region ahead of the crack. Although the exact values of the fracture strain in the two cases are different, if fracture mode remains unchanged, the tendency of the fracture strain associated with the grain size in both cases is identical. Therefore, the

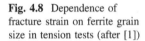

Fig. 4.8 Dependence of fracture strain on ferrite grain size in tension tests (after [1])

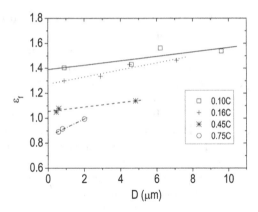

critical fracture strain in Fig. 4.3 also declines when ferrite grain size decreases from D_i to D_{i+1}. It should be noted that Fig. 4.3 merely shows the trend of the grain-refinement-induced variations in local stress, strain, σ_{cr}, and ε_{cr}, instead of the exact values.

Figure 4.3 schematically interprets the effect of grain refinement on the fracture mode. In the brittle fracture region (e.g., Case A), the multiplier effects of the elevated σ_{cr} and decreased local stress increase the difficulty of attaining the critical condition $\sigma_{yy} > \sigma_{cr}$. This means that in the brittle fracture region, grain refinement increases the fracture toughness of ferritic steels, which has been verified by Armstrong's work, in which fracture toughness K_{IC} has been shown to be proportional to $D^{-1/2}$ ($K_{IC} = 12.8 + 3.04D^{-1/2}$; K_{IC} in MPam$^{-1/2}$, D in mm) [10]. If the multiplier effects are sufficiently significant, the fracture mode can be changed from brittle to ductile. This type of transformation has been confirmed in ferritic steels with composition (0.10C or 0.16C)-0.3Si-1.45Mn (in wt%) [11].

In the ductile fracture region (e.g., Case C), for a given stress, a relatively large strain is produced for fine ferrite grains rather than for coarse ferrite grains. The increased strain and decreased critical fracture strain induced by the grain refinement are responsible for the occurrence of the ductile fracture with ease at a relatively low load, indicating that grain refinement has a negative impact on toughness. The experimental results that indicate that ferritic steels with fine grains have lower values of absorbed energy than those with coarse grains agree well with this conclusion [11].

For Case B, grain refinement changes the fracture mode from ductile/brittle fracture to ductile fracture.

The effect of the ferrite grain size on the fracture toughness has been schematically shown in Fig. 4.3. As for the actual fracture toughness, CTOD of ultra-fine-grained steels will be discussed in the following paragraphs.

Three different chemical compositions (steels D–F) given in Table 4.1 were selected for producing ultra-fine-grained steels [12]. Low values of welding crack sensitivity (P_{cm}) shows that steels D–F have strong resistance to cold cracking.

Table 4.1 Chemical composition of steels D-F and SM40B (in wt%) (after [12])

No.	C	Si	Mn	P	S	Sol Al	N	Nb	Ti	C_{eq}	P_{cm}
D	0.140	0.30	1.46	0.005	0.001	0.032	0.0013			0.38	0.223
E	0.095	0.30	1.45	0.005	0.001	0.032	0.0015			0.34	0.178
F	0.093	0.30	1.45	0.005	0.001	0.033	0.0018	0.016	0.007	0.33	0.176
SM490B	0.16	0.44	1.46	0.004	0.013					0.4	0.248

$C_{eq} = C + Mn/6 + (Cr + Mo + V)/5 + (Cu + Ni)/15$

$P_{cm} = C + Si/30 + (Mn + Cu + Cr)/20 + (Ni/60 + Mo)/15 + V/10 + 5B$

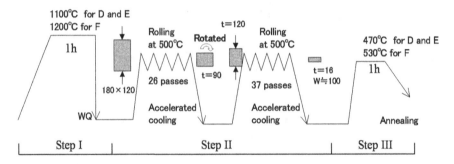

Fig. 4.9 Schematic illustration of the production processes (after [12])

Table 4.2 Mechanical properties of steels D-F and SM490B

Steel	YS (MRa)	TS (MRa)	YR	U.EI (%)
D	714	751	0.951	7.3
E	686	705	0.973	7.8
F	771	787	0.980	7.7
SM490B	340	524	0.649	18

Note YS yield strength, TS tensile strength, YR YS/TS, U.EI uniform elongation

For comparison, an SM490B steel with a thickness of 16 mm was used, whose chemical composition is also listed in Table 4.1. As can be seen from the table, the SM490B steel has almost the same chemical composition as that of steel D.

The steel plates D–F were produced by thermo-mechanical warm rolling with severe plastic deformation. In brief, the processes consisted of three steps, as shown in Fig. 4.9. In step I, slabs of steels D–F were prepared for rolling by heating in the furnace followed by water quenching (WQ). Step II is a rolling process in which slabs were first rolled in 26 passes and then rolled in 37 passes again after being rotated by 90°. Step III is an annealing process for improving the mechanical properties. The final cross-sectional size was approximately 100 mm (width) × 16 mm (thickness). The mechanical properties are given in Table 4.2. For comparison, the data of the SM490B steel with ferrite and pearlite microstructures are also listed.

The microstructure of all steels was observed by SEM, and the SEM micro-structures of steel D on the longitudinal and transverse sections are shown in Fig. 4.10a, b, respectively. It can be seen that steel D consists of ferrite and cementite particles. The morphology of the ferrite grains on the longitudinal and transverse sections was apparently different. On the transverse section, ferrite grains were fine (~ 1 μm) and equiaxial grains were dominant. In contrast, ferrite grains were considerably elongated on the longitudinal section, indicating that the longi-tudinal section consisted of a larger effective ferrite grain size than that in the

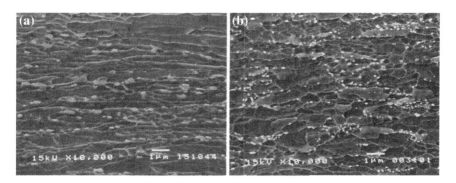

Fig. 4.10 SEM microstructure of steel D in **a** longitudinal and **b** transverse sections (after [12])

transverse section. The SEM observation showed that steels E and F had similar microstructures as that of steel D.

Three-point bending specimens with a fatigue pre-crack ($a_0/W \approx 0.5$) were prepared from the steel plates listed in Table 4.1 for measuring the CTOD of the steels D–F. The L- and T-direction specimens (fatigue pre-crack perpendicular and parallel to the rolling direction, respectively) were prepared with sizes 16 mm (B) × 30 mm (W) × 150 mm (L) and 16 mm (B) × 20 mm (W) × 100 mm (L), respectively. Two specimens were used for one condition. The CTOD tests were conducted at −40, −140, and −196 °C, with a span to width ratio (S/W) of 4. The value of CTOD was calculated by Eq. (4.8) proposed by ASTM E1290-99 [13]:

$$CTOD = \frac{K^2(1 - v^2)}{2\sigma_{ys}E} + \frac{r_p(W - a_0)}{r_p(W - a_0) + a_0 + z}V_p \tag{4.8}$$

where K is the stress intensity factor, V_p is the plastic component of the notch opening displacement, W is the effective width of test specimen, and a_0 is the average original crack length. In this paper, Poisson's ratio v is taken as 0.3, Young's modulus $E = 206{,}010$ MPa, plastic rotation factor $r_p = 0.44$, and distance of the knife edge measurement point from front face on specimen z = 0. Yield strength, σ_{ys}, at the temperature of interest was estimated from the yield strength, σ_{yso}, at room temperature by Eq. (4.9) for steels [14]:

$$\sigma_{ys} = \sigma_{yso} \times \exp\left[\left\{481.4 - 66.5 \times \ln \sigma_{ys0}\right\} \times \left\{\frac{1}{T + 273} - \frac{1}{293}\right\}\right] \tag{4.9}$$

where T is the test temperature in °C.

Figure 4.11 gives the load versus clip gage displacement for the L- and T-direction specimens of steel D. In order to clearly distinguish the curves, their zero points are shifted in Fig. 4.11. It is found that the load versus displacement curves in both Fig. 4.11a, b vary with temperature, and the variations in the curve shape due to temperature decrease in Fig. 4.11b are much more significant than

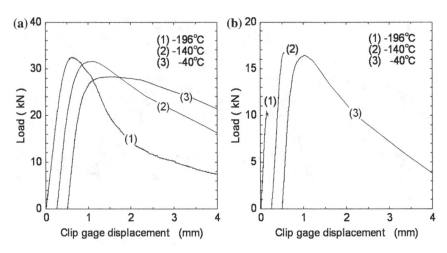

Fig. 4.11 Load against clip gage displacement of steel D. **a** L-direction and **b** T-direction specimen (after [12])

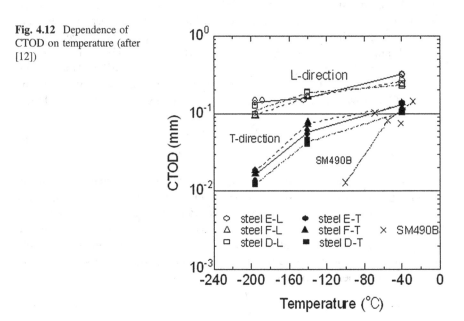

Fig. 4.12 Dependence of CTOD on temperature (after [12])

those in Fig. 4.11a. This implies that the fracture mode in the L- and T-direction specimens changed with temperature, particularly for specimens in the T-direction. Other steels also showed the same type of tendency.

The fracture surfaces of the CTOD specimens of steel D were observed by SEM. For specimens in the L-direction, the temperature for ductile fracture remained at −196 °C and the dimple size decreased with decrease in temperature. For specimens in the T-direction, steel D fractured in a fully ductile mode at −40 °C and then brittle fracture regions appeared and enlarged with decrease in temperature. This result is consistent with the insensitivity of fracture toughness to temperature for specimens in the L-direction shown in Fig. 4.11a.

The CTOD test results at the three temperatures are summarized in Fig. 4.12. The critical CTOD value decreased with temperature for the three types of ultra-fine-grained steels, irrespective of specimens in the L- and T-directions. However, the degrees of temperature dependence of fracture toughness differed. The critical CTOD of specimens in the T-direction decreased more strongly with temperature than those in the L-direction for a particular steel. Moreover, for a given temperature, the CTOD of specimens in the L-direction was apparently larger than that of specimens in the T-direction, which is attributable to the fracture modes described above.

As shown in Fig. 4.12, the CTOD values of the SM490B steel are also given. The SM490B sample obviously showed the worst low-temperature toughness among all the steels; moreover, its temperature dependence of fracture toughness was much more serious than that of the ultra-fine-grained steels, which indicates that the structure of ferrite/pearlite is much more sensitive than the ultrafine ferrite.

Figure 4.12 confirmed that refining the ferrite grain is an effective means of improving the fracture toughness; it can also lower the temperature sensitivity of the fracture toughness. Ultra-fine-grained steel has better low-temperature fracture toughness than hot-rolled steels with similar chemical composition.

References

1. H. Qiu, L.N. Wang, T. Hanamura, S. Torizuka, Physical interpretation of grain refinement-induced variation in fracture mode in ferritic steel. ISIJ Int. **53**, 382–384 (2013)
2. R.O. Ritchie, A.W. Thompson, On macroscopic and microscopic analyses for crack initiation and crack growth toughness in ductile alloys. Metall. Trans. A **16**, 233–248 (1985)
3. R.M. McMeeking, Finite deformation analysis of crack-tip opening in elastic-plastic materials and implications for fracture. J. Mech. Phys. Solids **25**, 357–381 (1977)
4. H. Qiu, L.N. Wang, T. Hanamura, S. Torizuka, Prediction of the work-hardening exponent for ultrafine-grained steels. Mater. Sci. Eng. A **536**, 269–272 (2012)
5. W.B. Morrison, The effect of grain size on the stress-strain relationship in low-carbon steel. Trans. ASM **59**, 824–846 (1966)
6. P. Antoine, S. Vandeputte, J.B. Vogt, Empirical model predicting the value of the strain-hardening exponent of a Ti-IF steel grade. Mater. Sci. Eng. A **433**, 55–63 (2006)
7. A. Ohmori, S. Torizuka, K. Nagai, Strain-hardening due to dispersed cementite for low carbon ultrafine-grained steels. ISIJ Int. **44**, 1063–1071 (2004)
8. A.H. Cottrell, Theory of brittle fracture in steel and similar metals. Trans. Met. Soc. AIME **212**, 192–202 (1958)
9. Armstrong, A.W.: In: Taplin, D.M.R. (ed.) Advances in research on the strength and fracture of materials, Fracture 1977, ICF4, Vol. 4, p. 83. Pergamon Press, Oxford (1977)
10. R.W. Armstrong, The (cleavage) strength of pre-cracked polycrystals. Eng. Fract. Mech. **28**, 529–538 (1987)

11. H. Qiu, R. Ito, K. Hiraoka, Role of grain size on the strength and ductile-brittle transition temperature in the dual-sized ferrite region of the heat-affected zone of ultra-fine grained steel. Mater. Sci. Eng. A **435–436**, 648–652 (2006)
12. H. Qiu, K. Enami, K. Hiraoka, K. Nagai, Y. Hagihara, Static fracture toughness of ultra-fine grained steels processed by thermo-mechanical warm rolling. J. Mater. Sci. **43**, 1910–1913 (2008)
13. ASTM E1290-99, Standard test method for crack-tip-opening displacement (CTOD) fracture toughness measurement (1999)
14. WES 2805-1997, Method of assessment for flaws in fusion welded joints with respect to brittle fracture and fatigue crack growth. The Japan Welding Engineering Society (1997)

Chapter 5
Summary

The main aim of this book is to establish fundamental concepts of using ultra-grain refining structures in future industrial applications. Apart from this, this book also introduced and discussed in detail some advanced steel techniques developed in the National Institute of Materials Science, Japan, in relation to exerting structural control, improvement of mechanical property, and mechanisms behind the phenomena of mechanical property improvements. The authors have focused particularly on tensile strength and toughness of advanced steels from both fundamental and engineering points of view. From the fundamental point of view, a unique approach of analysis based on the fracture surface energy using the effective grain size is employed to better understand the mechanisms behind property improvement. From the engineering point of view, the fracture toughness such as crack-tip-opening displacement (CTOD) in advanced steels was evaluated experimentally and compared to that of already established conventional steels.

The following are main conclusions obtained on the analysis of ultra-fine-grained steels that can potentially be used for the future development of advanced steels.

5.1 Tensile Properties

(1) During the annealing of submicron-grained ferrite/cementite steel with a heterogeneous and dense distribution of cementite particles, characteristic ferrite grain growth occurred in parallel with the Ostwald ripening of cementite particles. Predictive capability of the Hall-Petch relation between the hardness and average ferrite grain size was demonstrated, implying a significant potential for achieving hardening by means of grain refining.

(2) Lower yield stress, upper yield stress, and ultimate tensile stress exhibited monotonic relationships with the carbon content and volume fraction of the cementite particles. The true stress increased with increase in carbon content, while strain-hardening rate increases with increasing carbon content only up to 0.3 wt% C, after which the strain-hardening rate remained almost constant

© National Institute for Materials Science, Japan. Published by Springer Japan 2014   63
T. Hanamura and H. Qiu, *Analysis of Fracture Toughness Mechanism in Ultra-fine-grained Steels*, NIMS Monographs, DOI 10.1007/978-4-431-54499-9_5

even with further increase in the carbon content. This tendency of strain-hardening is well reflected in a similar change in the uniform elongation.

5.2 DBTT and Effective Grain Size

(1) The relationship between the effective grain size (d_{EFF}) and ductile-to-brittle transition temperature (DBTT) in the impact tests revealed that in the ultra-fine-grained ferrite/cementite (UGF/C) belongs to the same group composed of quenched (Q) and quench-and-tempered (QT) microstructures, while ferrite/pearlite (F/P) belonged to a different group. The estimated fracture stress showed that ultra-fine-grained ferrite/cementite microstructure exhibits the highest fracture stress among the four microstructures. The excellent toughness of the ultrafine ferrite/cementite steel is attributable to its characteristically small d_{EFF} and high surface energy of fracture, as compared to that exhibited by the other steel structures.

(2) Low absorbed energy with a ductile dimple fracture in the lower shelf region is a characteristic feature of the ultra-fine-grained ferrite/cementite (UGF/C) steel. In the ultra-fine-grained steel, there existed a transition from an energy-absorbent ductile mode to an energy-absorbent brittle mode in impact tests, where some dense and small dimples were observed in the lower shelf energy region.

5.3 Crack Tip Opening Displacement

(1) CTOD values of the SM490B steel showed the worst low-temperature toughness among all tested steels; moreover, its temperature dependence of the fracture toughness was more sensitive than that of the ultra-fine-grained steels, indicating that the structure of F/P was more vulnerable to stress than ultra-fine ferrite.

(2) Refining ferrite grains is an effective means to improve the fracture toughness; it can simultaneously lower the temperature sensitivity of fracture toughness. Ultra-fine-grained steel has better low-temperature fracture toughness than hot-rolled large grained steels with similar chemical compositions.

The authors hope that this text will find suitable readership among scientists, engineers, and university graduate students who are willing to gain a better understanding of the current and future technologies of steels manufacture, including steel science and engineering in terms of tensile, toughness, and mechanical properties in advanced steels, particularly UFG steels, specifically for environment-friendly applications.